ALL *the* RIGHT MOVES

The Definitive Guide for Integrating Dental Team Ergonomics, Treatment Room Technology, Auxiliary Utilization & Office Design

ALL *the* RIGHT *MOVES*

The Definitive Guide for
*Integrating Dental Team Ergonomics,
Treatment Room Technology,
Auxiliary Utilization & Office Design*

Risa Pollack-Simon, CMC
Simon Says Seminars, inc., Scottsdale, AZ

Dental Economics

This book is not intended to prevent (or cure) all of the problems that may occur in the work environment. The author is not engaged in rendering medical or legal advice. If medical or legal advice is required, the services of a competent professional should be sought. If you suffer from physical or stress-related disorders or conditions, seek professional guidance before attempting exercises or modifying postures or positions.

Copyright 2002 by
PennWell Corporation
1421 South Sheridan Road
Tulsa, Oklahoma 74112
1-800-752-9764
sales@pennwell.com
www.pennwell.com
www.pennwell-store.com

Cover design by Clark Bell

Library of Congress Info

All rights reserved. No part of this book may be reproduced, stored in a retrieval system, or transcribed in any form or by any means, electronic or mechanical, including photocopying or recording, without the prior permission of the publisher.

Printed in the United States of America

1 2 3 4 5 06 05 04 03 02

Goal

To take a deep breath of life and leave a legacy of balancing work and life with passion and comfort.

Dedication

This book is dedicated to my best friend and true partner, Wally Simon, who faithfully reminds me to take a deep breath of life, proclaim my rarity, and serve others with all the right moves.

Vision

It is my sincere intention that readers use this book as a hand to hold while choosing to move towards a distinctively higher level of performance, comfort, and health.

Table of Contents

Foreword ix

Step 1	**Let's Get Physical** 1	
	Posture and Stress	
Step 2	**Understanding the Human Body and Dem Bones!** 3	
	The Musculoskeletal System	
Step 3	**Prevention and Intervention** 11	
	Signs, Signals, and Healing	
Step 4	**Posture Perfect** 15	
	Achieving Balance and Comfort	
Step 5	**To See, or Not to See** 21	
	Optimizing Magnification	
Step 6	**What's Your Team's Position?** 27	
	Orchestrating the Four-Handed Team	
Step 7	**Environmental Space Factors** 33	
	Designing Work Environments	
Step 8	**Clinical Equipment** 37	
	Common Design Flaws	
Step 9	**Administrative Workspace** 49	
	Ergonomics on the "Front Line"	
Step 10	**The Master Design Plan** 53	
	Designing Your Dental Environment	

ALL *the* RIGHT *MOVES*

Step 11	**Integrating Technology**................................69	
	Networking the Clinical Environment	
Step 12	**That's a Stretch**...79	
	Conditioning Your Body	
Addendum	**The Occupational Safety and Health Administration** 87	
	Doing the Right Thing	

Bibliography 95

Glossary 99

Resources 103

About the Author 105

Index..107

Foreword

Welcome to the world of *ergonomics*—the science that studies musculoskeletal health and human performance of the worker on the job.

Workers compromise their musculoskeletal health when there is a mismatch between the physical capacities of the worker and the physical requirements of the job. When such a mismatch occurs, the worker risks developing musculoskeletal disorders—ailments of the muscles, nerves, and bone structures.

These disorders currently represent the nation's most serious occupational health hazard.

According to the Occupational Safety and Health Administration (OSHA), job hazards are costing workers their health and costing our economy a staggering amount—between $45 and $60 billion in worker's compensation, medical claims, and lost workdays annually.

Proactive ergonomic approaches, such as those outlined in this practice guide, are believed to prevent more than 300,000 musculoskeletal disorders each year. To that end, OSHA believes that ergonomic applications in the workplace provide American workers with a golden opportunity to proactively improve safety and performance on the job. Nonetheless, documented cases of musculoskeletal disorders continue to increase in dentistry. If employers and employees continue to ignore musculoskeletal disorders, they could end up carving a predictable path towards early retirement. Those who have been forced into early retirement have said that the emotional pain experienced in leaving the profession was oftentimes worse than the actual physical pain suffered on the job.

At Simon Says Seminars, inc., it has become our mission and professional responsibility to help dental professionals reduce these risks by teaching them how to provide efficient, high-quality patient care *with less musculoskeletal stress*.

You, too, can achieve this goal by simply becoming more aware of your posture and movements. The following tips will help you identify signs that indicate significant stress factors.

- Observe your habits, and continually monitor how often there is a mismatch between your physical capacities and the physical requirements of the tasks you perform

ALL *the* RIGHT *MOVES*

- Be cognizant of how often you find yourself leaning forward (or to the side) to see your work field
- Notice how often you have to break your concentration to retrieve items away from your immediate workstation
- Evaluate your side cabinets, dental units, and equipment to see if they are within arms' reach
- Track how often you run behind, lose track of time, or run over schedule

If you find that you habitually fall into these inefficient traps, you most likely are experiencing musculoskeletal stress and strain throughout the workday.

The actual magnitude of musculoskeletal disorders is not understood in dentistry because most individuals do not seek treatment or report symptoms until the symptoms worsen. Oftentimes, it is not until the pain becomes debilitating and affected individuals are completely unable to continue the practice of dentistry, that musculoskeletal symptoms are reported.

As you observe your habits and you choose to conscientiously move towards greater efficiency, balance, and comfort—we invite you to embrace the coach within the pages of this book to help you achieve comfort, safety, and health on the job.

STEP ONE

Let's Get Physical

"Ergonomic awareness is like a little angel that lifts us up into balanced posture when our own wings have trouble remembering."
—Risa Pollack-Simon

As we look at both past and current trends in dentistry, we can see a pattern of awkward postures and musculoskeletal stress and strain on behalf of dental providers and their auxiliary staff. Many of these risk factors are due in part to the constant desire to *get as close as possible to the work site to see better.*

Now add the challenge of positioning the patient into a fully reclined position, and you'll understand why dental care providers have relied on unbalanced postures.

The mere act of accessing the oral cavity when the patient's forehead, nose, and cheekbones are in the clinician's direct line of sight, clearly demonstrates why the view is obstructed and explains why poor postures prevail.

Historically, these poor postures have been used as survival tactics. In order to compensate for the interference, individuals would flex forward at the trunk or hold the shoulder of their dominant hand higher than the non-dominant side. The combination of these unsupported postures, excessive movements, and repetitive motions explains the evolution of musculoskeletal disorders in dentistry.

Attitude counts!

"A person will be just about as happy as they make up their mind to be."
—Abraham Lincoln

While physical risks seem to have the greatest impact on the musculoskeletal system, behavioral traits can also exacerbate stress levels considerably.

For example, it is not uncommon for dental professionals to feel overwhelmed with daily responsibilities as they constantly struggle to perfect their craft. This is particularly true of individuals who border on compulsive behavior, or who have an obsession with their own standards of excellence. Consequently, these unrealistic expectations compound existing stress and potentially compromise the health and well being of the individual.

Hence, *stress associated with performing dental care does not stem from the physical strain alone, but rather the combination of both emotional and physical factors.*

An honest evaluation of these human components can provide an accurate analysis and ensure that appropriate interventions are employed.

Stress Management

"There is a better way for everything. Find it."
—Thomas Edison

The willingness to learn time management techniques and master the art of delegation will empower employees and reduce stress considerably.

Most managers hesitate to delegate for fear of losing control or are concerned that the quality of the work will be inferior to what they could do on their own.

The need to be in control, combined with concern for good performance, can create more damaging results from burnout or exhaustion. This is especially true for those who "take it all on" or needlessly worry about the "to do" list and unrealistic deadlines.

STEP *TWO*

Understanding the Human Body and Dem Bones!

"The body is a miracle of optics, mechanics, chemistry, and electronics."
— B.J. Chang

The body is a fascinating machine. The brain, spinal cord, and peripheral nerves regulate all functions of the body and together are the source of all power. Nerves are hypersensitive and reactive to contact, and therefore must be protected. Hence, the primary objective of the spinal vertebrae is to protect spinal nerves, and the primary objective of the skull is to protect the brain from injury.

- *Vertebrae* are masterfully aligned to allow motion with their protective bony wall. In addition, each vertebra is separated by a cylindrical-shaped disc.
- *Discs* hold portions of the vertebrae together and provide shock absorption when the body is in a standing or sitting posture.
- *Spinal nerves* exit through openings between the vertebrae and stimulate muscles. The spinal bones are then anchored together by ligaments.
- *Muscles* help move the bones of the entire body as they contract and relax.
- *Tendons* are designed to further anchor the bones for extra leverage.

ALL *the* **RIGHT** *MOVES*

In addition to these key human components, the body also requires limbs for movement. The feet and legs provide support and balance. The legs are attached through the hip sockets and the pelvis to provide movement. The pelvis, combined with the sacral bone, is masterfully designed as the core foundation for the spine.

Following the universal laws of gravity, the human body relies on balance for spinal efficiency. Ideally, the upper body is balanced over the legs, with the spine perpendicular to the shoulders.

In dentistry, this position is an aberration. This in part is due to the fact that most clinicians move from balanced postures to extremely unbalanced postures to access the oral cavity.

The body in its infinite wisdom has a way of adapting to imbalances by twisting and turning other body parts to compensate for each deviation. This may cause the vertebrae to become lodged into abnormal positions or create restrictions on motion, resulting in nerve interference.

This process is also known as a *subluxation*, which can also become the source of spinal weakness, instability, and muscle spasm. These interferences prevent the body from functioning properly. In addition to associated stress, strain, and pain, the clinician's productivity can be hindered greatly, affecting practice growth considerably.

Checks and balances

The spine has four C-shaped curves, which can be seen more easily in a side view of the human body (Fig. 2-1). The first curve is located at the tailbone with the back of the C facing backward towards 9 o'clock; the second curve is in the lower back (lumbar area) with the back of the C facing forward towards 3 o'clock; the third curve is in the upper spine (thoracic area) with the back of the C facing backward towards 9 o'clock, and the last curve is in the neck (cervical area), with the back of the C facing forward towards 3 o'clock.

The forward curves *counterbalance* the backward curves, which allow the upper body to stay balanced over the center of gravity.

The goal is to maintain their natural connected C-shaped configuration to provide shock absorption and to reduce stress and strain on vulnerable areas of the back and neck, as well as abnormal wear or nerve irritation.

A well-designed and supportive chair can assist in obtaining this goal, providing the support features are utilized, and seated postures are balanced

Understanding the Human Body and Dem Bones!

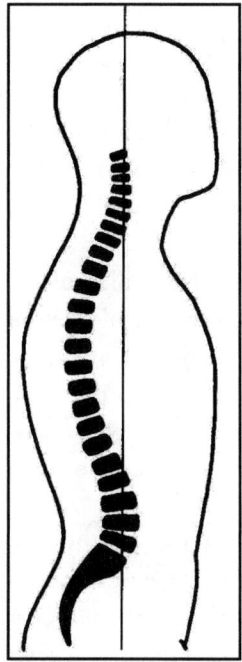

Fig. 2-1 Side View of the Human Spine. Reprinted with permission from *Sitting on the Job* by Scott Donkin, D.C.

over the center of gravity. We'll discuss this at length in succeeding chapters.

Body sculpting

Over time, poor operating positions will develop what is called tissue memory. The ligaments and other connective tissues that hold our bones together have an elastic spring-like property and a plastic, putty-like character. The elastic property will stretch and spring back into its original position.

Conversely, plastic properties have a tendency to remain at the position of greatest tension where they are held. Over time, the positions maintained the longest are the ones that sculpt your body postures. This is more readily seen on individuals who have observable abnormalities, such as a hunchback.

Move and groove

The joints and connective tissues—held together by ligaments—establish one's limited range of motion. Movements outside that set range can produce a dislocation or sprain, not to mention pain!

Conversely, if joints use less than their optimum range, over time they will end up restricting one's set range considerably. Therefore, it is just as important not to overwork your joints, as it is to not underwork your range of motion.

Pain, Pain Go Away...

"Everyone will experience the consequences of his own acts."
—Harry Browne

Pain, stress, and fatigue are symptomatic—red flags that a problem exists and should not be ignored. One does not have to suffer injury from a specific incident, such as heavy lifting, to experience these symptoms. In fact, in dentistry, it is much more common to experience less forceful pressure applied to body parts over a long period of time or through repetitious motions rather than to suffer an actual incident. These forces can gradually produce abnormal changes in the structures of the body, thus obscuring key signs of trauma. Consequently, this lag time dissuades a proactive approach towards appropriate actions that will remedy the problem.

Waiting for clinical symptoms to appear can exacerbate microscopic structural changes already developed, thereby furthering the musculoskeletal disease process. It should also be noted that the prognosis for healing can be proportionate to the length of time and severity of the injury. To that end, the longer an individual sustains stress factors, the greater the degree of tissue damage. Consequently, techniques and philosophies centered around prevention become key wellness strategies.

The choice of wellness, without pre-pathogenic clinical signs, however, becomes more challenging—*even for the most committed.* Hence, constant awareness and discipline must be employed. *Remember, the overall goal is not to exceed the structural limitations of the muscles, tendons, nerves, and joint structures.* When structural limitations are exceeded, a healing process must be employed as soon as possible, as healing is more responsive at the initial onset of signs and symptoms.

Healing involves the reduction of inflammation and the production of viable collagen fibers for adhesion. It is believed that healing must occur within the first three months to prevent re-injury. If a chronic cycle of micro-ruptures, inflammation, and/or painful adhesions occur, the affected individual can be more vulnerable to re-injury. Chronic sufferers may need to seek rehabilitation or physical occupational therapy.

Hand and Wrist Disorders

Carpal tunnel syndrome (CTS) is a nerve entrapment disorder that affects the median nerve as it passes through the carpal tunnel region of the wrist. Nerve dysfunction is generally due to chronic pressure induced by over-flexing and extending of wrist postures and repetitive motions.

Understanding the Human Body and Dem Bones!

In the carpal tunnel area, you will find eight carpal bones of the wrist along the floor, the *flexor retinaculum* along the roof and sides, and nine finger flexor tendons along with the median nerve traveling through the tunnel. Trauma occurs when excessive pressure is applied to the median nerve from repetitive activities (such as flexing/extending wrist postures while using vigorous pinch forces). Swelling occurs when the nine finger flexor tendon sheathes are overused (known as flexor tenosynovitis), which places pressure on the median nerve.

Symptoms of CTS include:

- paresthesia
- uncoordinated fingers
- rapid onset of hand fatigue
- hand weakness

Risks can be predisposed by many factors, however, chronic repetition of hand and wrist movements has been found to be the most debilitating, particularly when tension is also placed on the thenar and hypothenar eminence (the fleshy elevation of the palm).

Treatment options include changing patterns of hand use, keeping the wrist splinted at night, and the use of anti-inflammatory medication. If, however, surgery is required, a new method of surgery using endoscopic release through a one-centimeter incision is now available, according to the Mayo Clinic. This type of surgery can significantly reduce the recovery time—so much that procedures can be performed on both hands at the same time.

Other Factors

It is believed that ambidextrous gloves can exacerbate hand vulnerabilities. Conversely, the contours found in a right-left fitted glove support the natural pinch postures of the operator's hands during dental procedures due to the correct design of thumb placement. In addition, larger sized instruments, contoured handles, and silicone instrument sheaths can also reduce pinch-grip strain.

ALL *the* RIGHT *MOVES*

Rotator cuff tendinitis

Working with the elbow closer to the shoulder is known to cause rotator cuff irritation. Rotator cuff tendinitis is an inflammatory tendonopathy that most often involves the supraspinatus tendon (the tendon that helps lift the arm overhead), or any combination of the four rotator cuff tendons of the shoulder. The abuse of the proximal upper extremity region can also result in rotator cuff tendinitis. This can be seen more often when an individual has been treated for wrist therapy and is wearing a splint, which requires the shoulder to be used in an unnatural way to compensate for the loss of wrist motion.

Symptoms, which can be severe at night, include an aching discomfort of the shoulder, weakness, and pain. Diagnosis of rotator cuff tendinitis is usually made upon physical examination, sometimes requiring an MRI for confirmation. Treatment can range from modifying activities that exacerbate the problem to steroid injections.

Lateral epicondylitis

Lateral epicondylitis, more commonly known as tennis elbow, is caused by chronic repetitive wrist extension. This is seen more often when the wrist extensors are contracting against an opposing force, such as when a hygienist's wrist is extended outside neutral position during repetitive tasks.

Lateral epicondylitis symptoms include pain, especially during wrist extensions. For more advanced cases, the pain occurs at rest. Diagnosis can be made with a physical examination and is treated by modifying activities that exacerbate the problem. In addition, a counterforce brace, wrist splint, corticosteroid injection, or physical therapy may be required.

Medial epicondylitis

Medial epicondylitis, more commonly known as golfer's elbow, is caused by repetitive wrist flexion, particularly when combined with forceful muscle contraction against resistance.

Medial epicondylitis is seen as an inflammation of tendinopathy of the wrist flexor tendons, at their origin on the medial epicodyle of the elbow. Diagnosis is generally determined with physical examination. Symptoms include pain upon palpation or during wrist flexion. Treatment involves modifying activities that exacerbate the problem. In addition, a counterforce brace, wrist splint, corticosteroid injection, or physical therapy may be required.

De Quervain's tenosynovitis

Pain associated with De Quervain's tenosynovitis or De Quervain's disease is diagnosed when there is an inflammation of the thumb extendors and abductor tendons. De Quervain's syndrome occurs most often when bending the wrist in the direction of the fifth digit (known as an ulnar deviation). Diagnosis is made upon physical examination of the wrist in ulnar deviation while the patient makes a fist around the thumb. During this test, the patient will typically complain of intense pain at the site of the affected thumb tendons.

Symptoms of De Quervain's disease include pain along the base of the thumb, which increases with ulnar deviation. On occasion, the pain can migrate to the thumb extensor tendons and muscles in the forearm region.

Massage Therapy

Cumulative effects of daily stress can cause symptoms of various levels of discomfort. While massage therapy may not be the answer that solves the problem, it can be considered a very therapeutic form of healing. It has been stated that massage can benefit such conditions as muscle spasm, pain, spinal deviations (e.g., lordosis and scoliosis), soreness, headaches, and tension-related respiratory disorders, and it can even aid in reducing swelling, improving posture, and improving range of motion.

Massage works by releasing tension and increasing the rate of blood flow. This provides a gentle stretching action to connective tissues that surround muscles. Massage can also break up muscular waste deposits and stimulate circulation. Metabolic waste can accumulate and form painful knots. Massage can often times eliminate this tension and relax the body by easing stress away.

It is not a secret that dental professionals are exposed to a variety of upper extremity risks as a result of the tasks they perform in small, restricted spaces. It therefore becomes the ongoing responsibility of educators and trainers to increase the awareness of work-related musculoskeletal risks while applying preventive objectives in the work environment.

STEP THREE
Prevention and Intervention

"You cannot do wrong without suffering wrong."
— Ralph Waldo Emerson

As previously discussed, optimal human function is predicated by the ability to not exceed the structural limitations of one's muscles, tendons, nerves, and joint structures. To achieve this goal, ergonomic practices must be employed, and the onset of pain must alert the affected individual immediately that something is off balance.

As we have seen, when these signals are ignored, structural changes can occur, which can severely damage tissue and cause varied degrees of musculoskeletal disease over time. Hence, the longer the symptoms are ignored, the more severe the damage.

Intervention efforts can reduce inflammation and assist in the healing process when implemented in the early stages of disease development. Healing is critically important, as the process involves the laying down of viable collagen (scar tissue) over the injured area. If healing does not occur at the initial stage of soft tissue damage (typically seen in approximately 90 days), chronic disease may develop, involving a debilitating cycle of increased inflammation and re-injury.

There are three types of intervention that can reduce the magnitude of risk factors:

- Engineering Controls
- Work Practice Controls
- Administrative Controls

ALL *the* RIGHT *MOVES*

Engineering Controls are designed to reduce or eliminate exposure to musculoskeletal disorder risk factors. Engineering controls may include physical design changes, or equipment and instrument changes. For example, when equipment is not adjustable to fit the size and shape of the worker performing the task, the equipment should be replaced with an engineering control that is adjustable. Examples of engineering controls can include:

- Telephone headsets
- Keyboard drawers
- Document holders
- Touch screen monitors
- Adjustable extension arms
- Adjustable monitor mounting devices
- Screen filters
- Instrument cassettes
- Ergonomically designed instruments & handpieces
- Foot rests
- Flexible delivery systems
- Magnification & illumination
- Adjustable stools
- Fully reclinable patient chairs, with thin backs
- Recessed ultrasonic units
- Counter top grommets

Work Practice Controls are intended to modify behavior. They are used as a set of disciplines to modify the way tasks are performed. For example, when a worker is performing a task with increased risk, work practice controls are employed to minimize musculoskeletal risk factors. Examples of work practice controls can include:

- Positioning the keyboard so that the forearms are parallel to the floor
- Positioning documents to be entered into the computer at the same height as the computer monitor
- Adjusting the extension arms to ease access to the screen or keyboard
- Offering incentives for modifying tasks that exacerbate MSDs

Prevention and Intervention

- Sitting perpendicular to a window or using monitor screens to reduce eyestrain from potential glare
- Adjusting the seat pan, arm rests, and lumbar support of operating stools to achieve neutral postures
- Use proper positioning and posture guidelines to minimize stress and strain to the body
- Reclining the patient into a supine position to avoid flexing at the trunk
- Optimizing the use of four-handed dentistry with trained auxiliaries to minimize time and motion
- Magnifying and illuminating the field of view to minimize eyestrain

Administrative Controls deal with time management. They are intended to be used to prevent a worker from over-working the body and causing musculoskeletal risks. Administrative controls typically come into play with individuals who perform repetitive tasks or in high-volume practices that see a great deal of patients, or in practices that work into their lunch hours or past quitting time. Administrative controls can include:

- Limiting the amount of time a worker can perform a repetitive task without breaks
- Scheduling micro-breaks for stretching the body and resting the eyes
- Alternating work schedules to vary the duration of time spent on difficult tasks or procedures
- Sculpting work schedules for greater difficulty around the hours of peak performance

STEP *FOUR*

Posture Perfect

"If you care enough for a result, you will almost certainly attain it."
— **William James**

In an ideal home position, the worker is seated in what is known as a neutral posture. A neutral posture can be established by sitting upright with weight evenly distributed. Legs should be separated with feet flat on the floor. The back should be pressed against the back of the chair for lumbar support, and hips should lean forward to rotate the pelvis backward toward 9 o'clock. In this position, the "sit bones" should be palpable and the trunk should be reclined approximately 100-110 degrees (90 degrees is completely upright).

While providing dental care, we see the tendency to deviate the neutral position. Examples of such deviations include forward trunk flexion, cervical flexion, twisted torso, wrist deviations, and forceful pinch grips. These poor postures can lead to aches and pains, numbness, and tingling—all of which can accelerate the onset of cumulative trauma disorder.

Cumulative trauma disorder (CTD) is defined as a condition that develops due to repetitive tissue micro-trauma exceeding the adaptive capacity of normal tissue. The upper extremities are particularly prone to CTDs when performing repetitive tasks with poor posture. While the risk factors are relatively high, the painful conditions could be decreased with behavioral modification to postural habits in clinical practice.

ALL *the* RIGHT *MOVES*

Depending on the requirements of the job, healthcare workers can develop a number of disorders when structural limitations are exceeded.

Sitting on the job

"A well designed chair cannot achieve its maximum effect unless you, the user, understand how to make it fit your body and use it in that manner for which it was designed."
Reprinted with permission from *Sitting on the Job*
—Scott W. Donkin, D.C.

A metamorphosis took place when *sit-down* dentistry evolved more than 45 years ago. While obvious benefits in operator comfort were immediately acknowledged, an innocent ignorance emerged regarding seated postures.

The lack of ergonomic applications in dentistry impeded one's ability to understand how proper postures and chair support could reduce musculoskeletal stress, otherwise seen in stand-up dentistry.

The initial objective of sit-down dentistry was to take stress-bearing weight off the spine, legs, and feet. However, stress issues remained as these new technologies were introduced.

Hence, the myth of musculoskeletal risks being associated exclusively with stand-up dentistry was dispelled.

With little, if any, importance placed on the needs of the worker in relation to seated postures, dental care providers willingly became satisfied with whatever stool was provided, with little regard for comfort or support.

To make matters worse, manufacturers promoted stool give-a-ways with operatory equipment purchases—which further dissuaded the potential owner from scrutinizing their effectiveness.

For years the patient's comfort continued to be foremost, with the level of comfort for doctors, hygienists, and assistants holding a very low priority.

Fortunately, over the years, the evolution of four-handed dentistry (Fig. 4-1) and the awareness of time and motion efficiency opened the door for ergonomic efficiency and musculoskeletal health. Likewise, this new mindset created a stronger desire to take on work habits that promised greater efficiency while reducing musculoskeletal stress and strain.

Posture Perfect

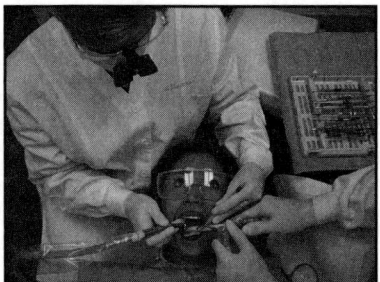

Fig. 4-1 Classic Four-handed Sit-Down Dentistry. *Courtesy of Dr. Jill Martenson, Oakland, CA*

The practice of four-handed, sit-down dentistry became the foundation surrounding ergonomic factors, which allowed clinicians to practice in balanced positions, using less time and motion.

Spinal balance

As we look back at a side view of a healthy human body, we are reminded of the four curves in the spine. The first curve is seen as a backward curve at the tailbone, which is followed by a forward curve at the lower spine, a backward curve at the upper spine, and finally, a forward curve at the neck.

Remember, the two forward curves counterbalance the two backward curves, which allows the entire trunk to remain balanced over the center of gravity. Balance is important while performing tasks, particularly in a seated position. Therefore, the design and utilization of the worker's chair become critical parts of this support formula.

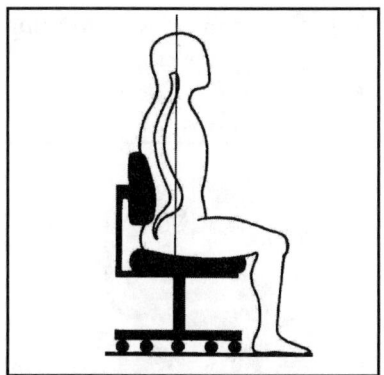

Fig. 4-2 Your chair should fit the length, size, and contour of your body for maximum support and comfort.

Your chair should fit the length, size, and contour of your body for maximum support and comfort. Your thighs should be at a right angle to your back. When you sit back into the seat pan, you should sit back as far as possible to take maximum advantage of the lumbar support feature of the chair (Fig. 4-2).

If not seated back and snug into the seat pan, you will have a tendency to lean forward. This slumping posture increases pressure on discs and makes the muscles and ligaments of your back work harder.

The assistant's stool should provide five mobile casters and a ring or platform at the base (to support the feet), with a wide seat pan (adjustable in height and tilt), and an adjustable lumbar support (Fig. 4-3).

ALL *the* RIGHT *MOVES*

The operator's stool (doctor, hygienist, or expanded duty auxiliary) should also have five casters at the base, a wide seat pan (adjustable in height and tilt), an adjustable lumbar support, and adjustable forearm rests (Fig. 4-4).

The backrest should be centered at the cross section where the forward curve of the lower back meets the backward curve of the middle back. This will help keep the shoulders upright and reduce weight and pressure on the lower back.

The height of the chair is determined by the position of the forearms in relation to the upper arms. The forearms should be at a 75 to 90-degree angle to your upper arms.

When forearm rests are provided and properly utilized, the provider can take up to 12% of the stress-bearing body weight off the spine, which further reduces stress and strain to upper back and shoulders. Armrests should be adjustable and removable. The key objective is to adopt a position that allows optimum access, visibility, comfort, and control. Muscles should be relaxed and well balanced (with the exception of those muscles performing the task).

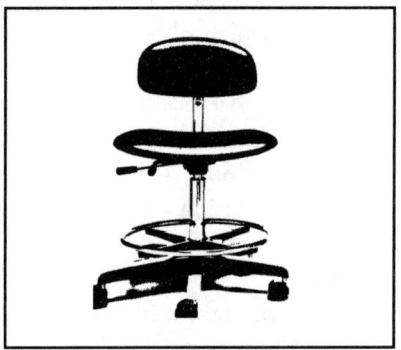

Fig. 4-3 The assistant's stool should provide five mobile casters and a ring or platform at the base (to support the feet), with a wide seat pan (adjustable in height and tilt), and an adjustable lumbar support. *Used with permission from* Link Ergonomics, Napa, CA: Assistant's Chair185ABR

Fig. 4-4 When forearm rests are provided and properly utilized, the provider can take up to 12% of the stress-bearing body weight off the spine, which further reduces stress and strain to upper back and shoulders. Armrests should be adjustable and removable. *Used with permission from* Link Ergonomics, Napa, CA: MicroSurgeon's Chair (MLS)

Don't let your focus on patient care distract you from these guidelines. As a result, good spinal balance and improved musculoskeletal health can enhance the quality of patient care—making these work habits a win-win for all.

New concepts in sitting postures

Research in the fields of sitting ergonomics and spinal biomechanics indicates that pain and stress (especially in the lower back area) result from sitting in positions where the spine cannot maintain its natural curvature. Likewise, repetitive muscle work can cause injury, particularly when the worker is sitting too low in relation to the workstation. New research shows that pain in the lower back, spine, shoulders, and neck increases with the amount of time a person is sitting.

One unique Australian saddle-style chair design appears to keep the pelvis balanced and stable, providing optimal spinal function and balance.

This new design supports the natural spinal curve by positioning a person to have his/her feet on the floor in a spread-leg stance. This position keeps the head directly over the feet, which produces a stable posture and supports free and controlled movements (Fig. 4-5).

The saddle therefore facilitates sitting at the correct height and produces a very finely tuned balance between shoulders, neck, and head, helping to improve ease of movement and the precision of hand-eye coordination. It also is believed that this upright posture can eliminate harmful compression of the thorax and abdomen, which, in turn optimizes respiratory and digestive function.

The most unique difference between the saddle-type chair and conventional chairs or stools is seen in the "open" leg position. The open-leg posture is said to enhance blood circulation and eases a person's ability to sit down or stand up from the chair, as opposed to conventional chairs.

If you do not ride horses (as does the founder of the Bambach® Saddle), you most likely will experience discomfort with the open-leg posture initially. This new seated position not only feels different—it also puts a fair amount of pressure on the pelvic area and therefore its use must be eased into your daily routine—not unlike the learning curve seen when wearing orthotics or magnified loops.

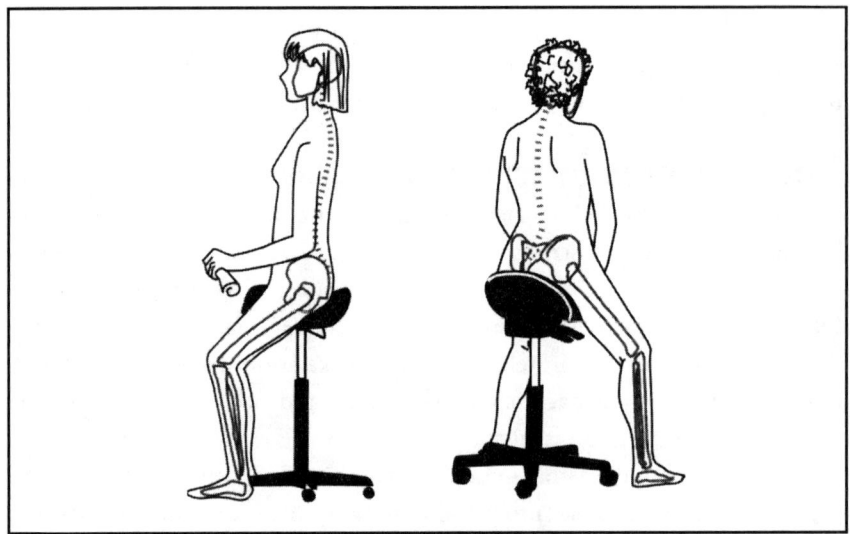

Fig. 4-5 Australian saddle-style chairs support the natural spinal curve by positioning the feet on the floor in a spread-leg stance. This position is believed to improve balance between shoulders, neck, and head, helping to improve ease of movement and hand-eye coordination. (Courtesy of Bombach Saddle Seat Pty Ltd.)

STEP *FIVE*

To See, or Not to See

"Seeing all possibilities can make what used to be just ordinary—quite extraordinary!"

— Risa Pollack-Simon

Magnification through properly designed surgical telescopes (known as loupes) and microscopes has proven to be a most beneficial aid to maintain neutral sitting postures. While the most obvious benefit of magnification is visual acuity, there is a hidden gem gained through improved posture from the working length requirement.

The working length is the distance from the work site to the operator's eyes. Therefore, the working length of the operator's scopes becomes a critical linking mechanism to ideal posture. The working distance should be adjusted to support ideal posture (approximately 12-16 inches). When the working distance is adjusted below that range, the operator can be forced to bend at the trunk.

The working distance should ideally be calibrated to ensure that the operator's neck, shoulders, and back are not over-extended (leaning back) or over-flexed (leaning forward). The depth of field must also be evaluated. The depth of field (Fig. 5-1) is the range of area in focus and is determined by the nearest and furthest extremes of distance from the surface of the operator's eye to the treatment site.

ALL *the* RIGHT *MOVES*

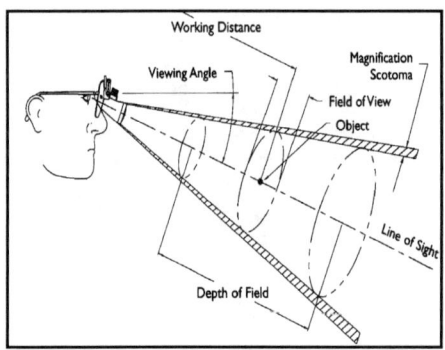

Fig. 5-1 The working distance should be adjusted to support ideal posture (approximately 12-16 inches). When the working distance is adjusted below that range, the operator can be forced to bend at the trunk. *Courtesy of SurgiTel/General Scientific*

The *angle of declination* plays a very important role in the calibration of magnification. The angle of declination is the degree in which the eyes are declined while operating. Ideal declination angles range from 15 to 44 degrees.

If the range of the scopes does not provide up to a 45-degree angle of declination, the operator will be forced to flex the neck downward to recreate an ideal angle of view. This over-flexed position puts unnecessary stress on the musculoskeletal system, causing neck and shoulder pain and potential musculoskeletal strain.

When evaluating the angle of declination, the operator must assume a working position and simulate the treatment environment. While maintaining an upright neutral sitting posture, with the patient fully reclined, the operator should attempt to view the maxillary anterior teeth without leaning forward or dropping the neck downward.

The ideal angle of declination, therefore, becomes the cross-section between balancing both extremes of eye comfort and neck comfort.

The convergence angle must also be set precisely to reduce eyestrain and improve resolution. The convergence angle is the pivotal angle aligning the two oculars, such that both oculars are pointing at the identical distance and angle. The manufacturer should precisely set convergence angles.

The power level of magnification is a personal decision, depending on the precision of the care you provide and your comfort level. Most clinicians desire higher powers for greater detail—however, it should be noted that higher powers also reduce your depth of field.

Likewise, the higher the power, the more challenging it will be to keep your field of view steady. As a rule of thumb, the higher the power, the more jolting minor movements will appear. In addition, the higher the power, the larger the scotoma (the blind zone between the peripheral unmagnified view and the center of magnification). Typically, most dental procedures can be performed between 2 and 4x.

Design

Surgical magnification can be purchased in two different styles, "through-the-lens" or the "flip-up."

Through-the-lens

Through-the-lens may be lighter in weight; however, they do not allow for any adjustment. Likewise, the eyeglass lenses tend to be thicker to firmly hold the telescopes, and the frames tend to be heavier to maintain the optical alignment of the telescopes.

Through-the-lens styles do not offer flexibility for multiple users. Furthermore, the user is inconvenienced when the prescription of the eyewear changes, as the lenses must be returned to the manufacturer for replacement—unlike the flip-up style.

Flip-up style

The flip-up style can adjust from a magnified view to a non-magnified view, without removing the eyewear from the operator's face (Fig. 5-2). Newer designs have added vertical adjustment capabilities in the declination angle for optimum posture with different procedures. Most flip-up scopes also allow interpupillary adjustment for multiple users.

Older style flip-up telescopes were heavier than through-the-lens models. Current technology has allowed some new designs of the flip-up style to be even lighter than through-the-lens scopes by using lighter frames and thinner glass lenses. Flip-up styles allow the user to change the prescription of eyewear at a location of choice—minimizing downtime considerably.

Microscopes

After a clinician experiences the benefits of loupes for a period of time, he or she may desire a higher magnification level. With higher magnification systems, the loupes become heavier and more cumbersome. The microscope

ALL *the* RIGHT *MOVES*

Fig. 5-2 The flip-up style can adjust from a magnified view to a non-magnified view, without removing the eyewear from the operator's face. *Courtesy of SurgiTel/General Scientific*

(Fig. 5-3) may offer a range of magnification that can serve all clinical needs, from the ability to diagnose at lower power of operation to a much higher power for inspection. In addition to offering multiple levels of magnification, the microscope also offers a built-in, coaxial, high-intensity light source.

This environment enables the operator to work in a focused fashion using fewer movements. It is believed that microscope users receive much more visual information about the treatment area than an operator with unaided eyes. This results in more precise neuro-motor skills—thus, procedures are more delicate, deliberate, and precise. A higher level of precision also allows the operator to function with greater efficiency, productivity, and stamina. In essence, the microscope is a unique piece of technology, in that by using it one can make all other technologies—including human performance—even better.

Illumination

"Keep your face to the sunshine and you cannot see the shadows."
—**Helen Keller**

As the field becomes more magnified, it tends to require more light or illumination. Illumination—in addition to overhead lights—can be added to the magnifiers in either a head mount or a direct mount on the spectacle frame or the telescope-mounting fixture. One of the key benefits of this coaxial illumination is to provide clinicians with shadow-free images.

Illumination is available in either fiber optic or direct lamp (halogen or xenon) styles. Typical fiber optic lamps (Fig. 5-4) offer a range of 2,000 to 35,000 lux (or 200 to 3,500 foot-candle). Direct halogen lights (Fig. 5-5) generate 3,000 to 10,000 lux (or 300 to 1,000 foot-candle). Although direct halogen lights cannot offer the intensity of light equal to that of fiber optic lights, they have been shown to be quite adequate for common restorative procedures.

Direct halogen lights also offer the advantages of cost savings and ease of mobility with their clip-on battery pack design. This eliminates the need to connect and disconnect the light source when moving from room to room.

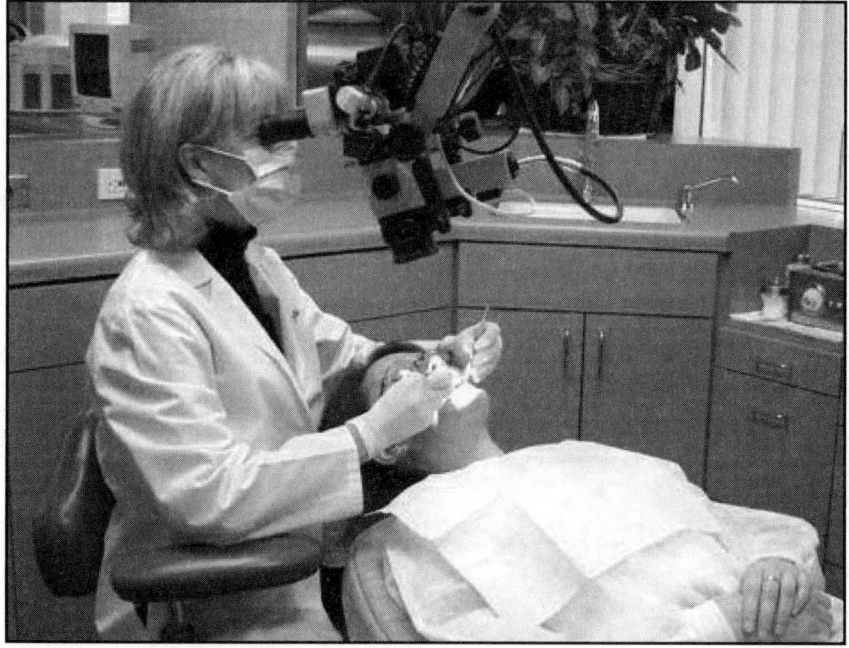

Fig. 5-3 Dr. Jacinthe Paquette performs restorative dentistry through a surgical microscope, which allows greater precision in care due to increased visibility from magnification and illumination. In addition to her private practice and international lectures, Dr. Paquette helps dentists learn how to use the microscope to reduce musculoskeletal stress, while enhancing treatment outcomes at the Newport Coast Oral Facial Institute in Newport Beach, CA. *Courtesy of Dr. Cherilyn Sheets*

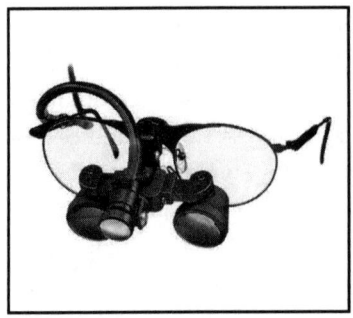 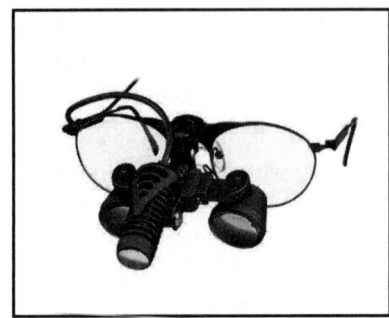

Figs. 5-4 and 5-5 Illumination is available in either fiber optic or direct lamp (halogen or xenon) styles. Typical fiber optic lamps (Fig. 5-4) offer a range of 2,000 to 35,000 lux (or 200 to 3,500 foot candle). On the other hand, direct halogen lights (Fig. 5-5) generate 3,000 to 10,000 lux (or 300 to 1,000 foot-candle). Although direct halogen lights cannot offer the intensity of light equal to that of fiber optic lights, they have been shown to be quite adequate for common restorative procedures. *Courtesy of SurgiTel/General Scientific*

STEP *SIX*

What's Your Team's Position?

"Position in dentistry is everything!"

— Risa Pollack-Simon

The position of the team can play a critical role in ergonomic practice. Proper posture, positioning, and the synchronistic movements of the team can provide an orchestrated procedure, thus saving time and conserving energy.

To best illustrate the ideal position for the restorative operating team, examine the clock face and compass in Figure 6-1. The clock face is positioned with 12 o'clock facing north and 6 o'clock facing south.

With the patient fully reclined, the right-handed operator is positioned facing SE with his/her left leg extended underneath the patient's chair and the right knee at a 90-degree angle, operating the rheostat, facing SW.

In four-handed dentistry, the goal is to optimize the use of the auxiliary's hands and eyes in an effort to minimize time and motion for the operator. This can be best accomplished by positioning the assistant so that support equipment (high-velocity suction and 3-way syringe), the work surface, and the dental unit (with handpieces) are as close as possible. This philosophy assumes that the assistant is the primary user and discourages operator retrieval, while still allowing the operator to have access when needed.

In this model, the assistant is responsible for instrument and handpiece exchanges, material preparation, rinsing and drying the field, tissue retraction, illumination, and oral evacuation.

27

ALL *the* **RIGHT** *MOVES*

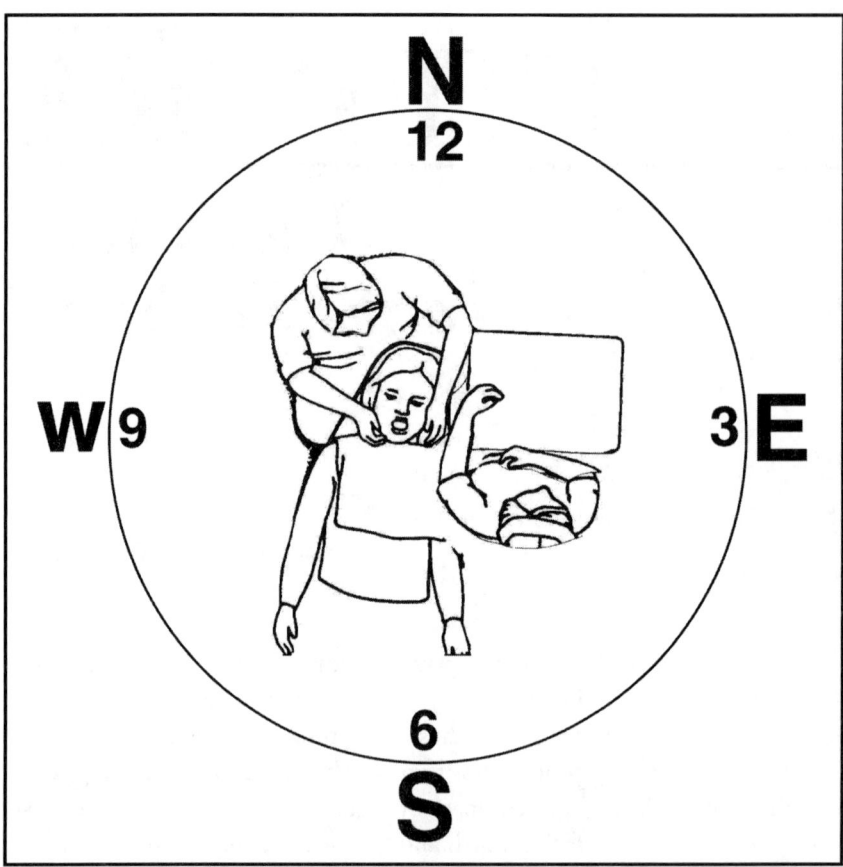

Fig. 6-1 The ideal position for the restorative operating team can be viewed on a clock face and compass.

To enhance access and reduce reach distances, the work surface must be positioned directly over the assistant's lap (Fig. 6-2). This will allow the assistant quick and easy access to the suction, air-water syringe, instrumentation, burrs, and handpieces. During procedures, the assistant maintains the air-water syringe in the left hand, with the high velocity evacuator (HVE) in the right hand to maintain a clear field when assisting a right-handed operator.

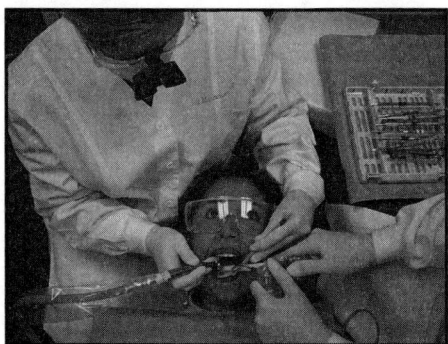

Fig. 6-2 To enhance access and reduce reach distances, the work surface must be positioned directly over the assistant's lap. *Courtesy of Dr. Jill Martenson, Oakland, CA*

When instruments or burrs must be changed, or materials need to be manipulated, the suction and air-water syringe are transferred into the left hand (or returned to the unit holder) to free the right hand for these required tasks.

When operating from dental units designed with the suction and air-water syringe mounted behind the assistant, new holding devices should be mounted at the front work surface. This will ease assistant access and minimize movement. Ideally, the suction should be mounted on the right side of the work surface with the air-water syringe on the left. (When assisting a left-handed operator, all of the aforementioned positioning is reversed.)

To maintain visual access to the operating site and protect tissue integrity of the lips, cheeks, and tongue, the operating team must manage soft tissue at all times. This can be accomplished by using the aspirator, mirror head, air-water syringe, cotton rolls, the back handle of the cotton forceps, or a combination of these items when necessary.

As a rule of thumb, the team member seated closest to the tissue requiring retraction is responsible for retraction of that area (Fig. 6-3). This prevents awkward and unbalanced postures, which could cause shoulder stress from elevated elbows and extended or raised arms.

During exchanges, the operator's eyes are fixed on the treatment site, and a class-one motion (involving the fingers of one hand only) is used to indicate that an exchange may be initiated. By using a class-one motion (a slight withdrawal of the instrument in use by the fingers only), the tip of the working end is placed in the assistant's line of view, which allows the instrument to clear potential contact with soft tissue during the exchange. Until the class-one signal is initiated, the assistant maintains the anticipated instrument in a "holding pattern" below the patient's chin.

ALL the RIGHT MOVES

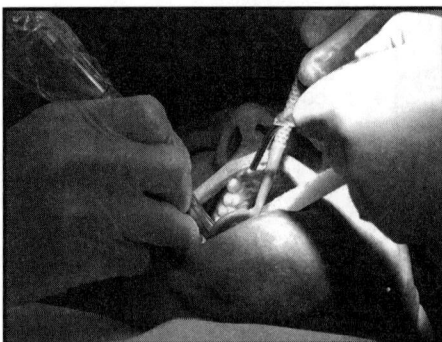

Fig. 6-3 As a general rule, the team member seated closest to the tissue requiring retraction is responsible for retraction of that area.

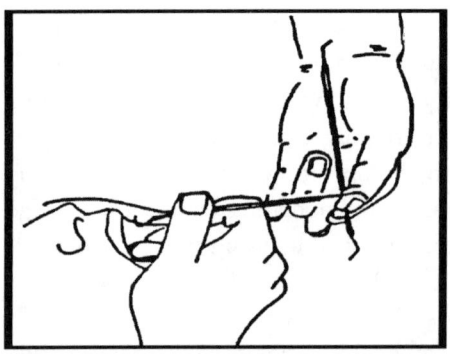

Fig. 6-4 When the assistant is ready to proceed with the transfer, the little finger contacts the operator's top knuckle of the index finger, and then proceeds down the top of the index finger. This will provide the operator a physical indicator that the instrument transfer is in process without the need for visual contact. A demonstration of instrument transfers can be viewed on *Smooth and Efficient Patient Care at the Chair,*" a training video. See: www.simonsaysseminars.com.

The holding pattern involves holding the instrument to be transferred in the midline of the assistant's palm. The palm should be parallel to the ceiling, with the little finger extended below the operator's right hand just below the patient's chin. When the assistant is ready to proceed with the transfer, the little finger contacts the operator's top knuckle of the index finger, and then proceeds down the top of the index finger. This will provide the operator a physical indicator that the instrument transfer is in process without the need for visual contact. (Fig. 6-4)

The next step involves retrieval of the old instrument with the little finger. This is accomplished by curling the little finger underneath the expired instrument, while the index finger and thumb position the new instrument into the operator's hand.

The instrument should always be positioned into the operator's hand within the operator's field of view—with the working end toward the treatment area. This is of particular importance when the operator is wearing magnification.

The goal is to position the new instrument into the operator's hand without the operator releasing fulcrums or adjusting the eyes away from the treatment site.

Patient positioning

When the patient is incorrectly positioned, the clinician's posture is greatly impacted. For example, when the patient is not fully reclined, the operator will need to flex the trunk and neck forward, which compromises musculoskeletal health and comfort.

To gain patient compliance with ideal chair positioning, it is recommended that the chair be brought back in stages to achieve a fully reclined position without objections. In addition, the assistant can verbally reassure the patient as he/she is reclining the chair, to further defuse fear and help the patient feel safe and in control.

Of equal importance, the patient's head should be positioned to offer maximum access and visibility. When the patient's head is improperly positioned, the operating team can be forced to lean from side to side or "wing" their elbows up and away from the sides of their body.

To avoid awkward postures and enhance visual access, the patient's head should be adjusted. For example, when treating the occlusal surface of any quadrant, rotate the patient's head toward the operator. However, when treating the buccal surfaces, rotate the patient's head toward the lingual aspect. Likewise, when treating the lingual surfaces, adjust the patient's head toward the buccal. These guidelines will bring each treatment surface closer into view and avoid unnecessary strain.

Considering that the patient is only in the chair for a short period of time (as it relates to a clinician's workday), it behooves one to take advantage of these guidelines designed to improve operating team comfort and performance.

Maxillary position

When treating the maxillary arch, the patient's chair should be reclined into a supine position. To improve access and visibility, the head should be positioned as far to the end of the headrest as possible. Additionally, an oblong pillow should be positioned under the patient's neck to support cervical vertebrae. This naturally brings the chin up and places the occlusal surface of the maxillary arch perpendicular to the floor (Fig. 6-5).

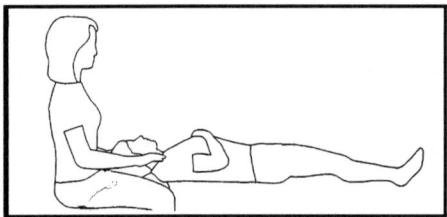

Fig. 6-5 When treating the maxillary arch, the patient's chair should be reclined into a supine position. To improve access and visibility, the head should be positioned as far up to the end of the headrest as possible, and an oblong pillow should be positioned under his/her neck. This will naturally bring the chin up, placing the occlusal surface of the maxillary arch perpendicular to the floor.

The operator's stool height should be adjusted so that the patient's head is level with the operator's elbows. In addition, the operator's shoulders should be relaxed, with arms resting at the sides of the body. Be certain that the elbows or shoulders are not raised above a neutral sitting posture.

Mandibular position

For the mandibular treatment areas, the patient's chair can be elevated 10 to 15 degrees upward from the fully supine position. This will position the mandible parallel to the operator's elbows, at elbow height. The patient can be asked to raise or lower the chin when needed.

Moreover, the patient's chair height should be low enough to align with the operator's elbows, yet not impinge on the operator's leg, which is extended under the patient's chair. Thin chair backs provide ease of obtaining these relationships.

STEP SEVEN

Environmental Space Factors

"Ergonomic environments support greater efficiency, productivity, and performance!"

— Risa Pollack-Simon

In my travels, consulting in dental offices nationwide, I became fascinated with the variance in dental office designs and modes of operation. I was equally intrigued with the amount of detail that was placed on decor and the lack of emphasis on functional layout as it relates to efficiency.

Consequently, I traveled the country searching for the most functional design. My goal was to identify and incorporate essential environmental factors that allowed clinicians to work smarter—*with less stress!*

I realized that in order to implement such a concept, I would first have to dispel the long-standing belief that stress and strain were necessary evils associated with the practice of dentistry. I began by coaching operating teams to "work smarter." I demonstrated how low back pain, tendonitis, nerve problems, and other forms of musculoskeletal injuries could be minimized—*if not eliminated.* I encouraged behavior modification in posture and team positioning. I coached dental teams on how to minimize body movements, maintain balanced posture, and reduce repetitive motion or forceful exertions and pinch forces.

I also modified spatial relations and support equipment for ease of access, and found that equipment was positioned far beyond one's natural reach (ideally set at a 20-inch radius).

In addition, I evaluated four-handed and six-handed dental auxiliary utilization and found that auxiliaries were not fully utilized—and that chair time was much greater than what was scheduled!

All in all, my findings revealed that assistants were underutilized, and doctors were spending too much time (and using too much motion) getting their own instruments and handpieces. Much of this was due to the design and position of the dental unit, instrument systems, and support cabinetry.

Job station analysis

Each workstation should be evaluated to ensure that it is designed to meet the needs of the worker. Extreme positions and extended reaching distances are red flags, along with static postures and muscle imbalances.

Risk factors include:

- unbalanced posture
- poorly designed dental equipment
- inaccessible treatment sites
- insufficient auxiliary utilization
- poor patient positioning

Inadequate lighting, excessive noise, and unhealthy temperatures have also been found to be significant risk factors in the workplace.

In an ideal ergonomic setting, all primary work surfaces and support equipment should be positioned directly over the assistant's lap with the air-water syringe and suction within arm's reach. Handpieces must also be accessible to the assistant to ensure that the assistant can retrieve or exchange handpieces or change burrs. Ideally, reach lengths should not exceed a radius of 20 inches.

When the dental unit is split into two separate systems, the handpieces are typically placed beyond the assistant's reach. In such a case, the operator must leave the operating field quite often, which interferes with levels of productivity and becomes a major distraction in concentration.

Environmental Space Factors

Frequency of use should also be taken into consideration. Support equipment and controls with high frequencies of use should be positioned at the optimum location without obstructing access to other controls. For example, the high-speed handpiece should be positioned closer than the slow-speed handpiece, and the high-velocity suction should be more accessible than the saliva ejector.

Legroom is also important, to obtain clearance underneath the work surface. To obtain leg and knee clearance, the thickness of the work surface should clear at least two feet at the foot and 16 inches at the knees. To achieve these dimensional requirements, equipment must be adjustable and allow adaptation to multiple users.

Lighting

Tasks that require high visual acuity and contrast sensitivity will function best at high levels of illumination. Intricate tasks should be illuminated at approximately 1,000 to 10,000 lux. Office tasks require low background lighting—approximately 300 to 700 lux—combined with non-glare, adjustable, task-directed lighting.

Noise

Noise levels should be monitored for safety. According to the Occupational Safety and Health Administration's standard (OSHA standard 29 CRF 1910.95), exposure to noise can lead to temporary or permanent deafness, tinnitus, paracusis, or speech misperceptions. It is believed that levels and duration of noise can directly and proportionately increase risk factors.

Strategic Plan

The application of these recommendations requires a strategic plan, a timeline for implementation, and a realistic budget analysis. A willingness to challenge traditional habits is also required. Taken together, these will provide an effective means of ensuring optimal team health and overall practice success.

As you apply these principles to your practice, you will begin to see exciting changes, such as increased efficiency and productivity. More importantly, you will have taken a positive step toward ensuring a longer, healthier—and more energized—practice lifestyle.

STEP EIGHT

Clinical Equipment

"Failure is only the opportunity to more intelligently begin again."
— **Henry Ford**

The National Institute for Occupational Safety and Health (NIOSH) has recorded more than 160 musculoskeletal and nervous system disorders, ranging from back and joint pain to tendonitis. These conditions can be partly ascribed to poor design of equipment, technical systems, and tasks.

In recognition of these concerns, California was the first state in the nation to finalize an ergonomic standard. As for the federal guidelines, and other state-run programs, individuals continue to challenge ergonomic regulations for lack of scientific data.

To that end—and until we collect additional statistics showing trends that support ergonomic interventions—worker's compensation and medical costs will most likely continue to increase, and the musculoskeletal health of the worker is almost sure to continue to decline.

Most employers do not like to acknowledge that the work environment and certain tasks they require employees to perform could be hazardous to the health of those employees. Much of this is due to Western society's persuasion and our perceptions of what is "bad" and what is "good." It has taken us many years to accept smoking, drinking, overeating, and personal stress as high-risk factors. It is time to add physical stress to that list to ensure the necessary action for behavioral changes.

ALL *the* RIGHT *MOVES*

As ergonomists continue to document musculoskeletal disease and stress as governing factors related to sick days taken and disability claims filed, it will become more challenging to look the other way.

While the occurrence of these disorders may not be as great in the dental profession as it is in others, a preventive approach must be applied, regardless of regulatory requirements.

Although the science of ergonomics may not be exact, methods to control known risk factors must be enacted to further prevent musculoskeletal disorders and ensure safe work environments.

According to the October 1994 issue of *Forbes* magazine, OSHA's entry into the ergonomic arena was not by chance, but rather by circumstance. The magazine was told that OSHA inspected and fined Pepperidge Farm $1 million, alleging that management forced employees on a cookie line to perform too many repetitive hand-motion tasks. Pepperidge Farm contested OSHA's actions and the case was dismissed. Nonetheless, it is said that the judge handling this case informed OSHA that the agency *could* enforce employers to comply with ergonomic standards, but only after it created and published guidelines based on scientific evidence. It was after that occurrence that draft regulations were developed.

The Department of Labor has made the guidelines available through the Internet to anyone who cares to download the document. (*Caution:* It runs more than 500 pages of regulatory text and voluminous supporting appendices!)

The standard, once finalized, would require employers to identify those jobs that pose musculoskeletal risks and implement a training program for identifying risks and preventing them. Likewise, the standard requires that improvement programs are put into place and that employees report incidents and receive medical follow-up and work restriction protection.

Work-related musculoskeletal risks are defined as activities that cause or aggravate symptoms of pain or discomfort. These symptoms are more commonly found in the neck, shoulders, hands, wrists, and lower back area. Stress-causing mental or physical exhaustion is also included.

It is believed that risk factors are often increased by either using vibrating tools, a fixed or awkward work posture, or the same motion pattern every few seconds.

Clinical Equipment

Since these conditions can be partly due to poor equipment design and/or the activities performed while operating such equipment, employers are encouraged to work with manufacturers and ergonomic consultants when developing job improvement programs.

According to OSHA, an employer's objective is to provide a safe work environment. With that in mind, a smart manufacturer's objective would be to network with employers and ergonomic consultants in an effort to design equipment and instrumentation around ergonomic safety principles.

I believe dentistry can benefit greatly from applied ergonomics. Aside from the desperately needed safety benefits, a practice can dramatically improve performance objectives through greater productivity.

As a management consulting firm specializing in time and motion efficiency, we are given many opportunities to monitor human performance and stress factors in dental offices nationwide. Over the years, we have gathered information and documented trends that appear to contribute to musculoskeletal disorders. We have also seen that these disorders and poor work habits contribute to lost time and lost production, which can significantly affect the quality of teamwork. Once these inefficiencies are identified and corrected, operating teams can begin to modify posture and work habits by adopting an improvement plan for maximum efficiency and comfort.

The Eight Major Flaws and How They Can Be Corrected

We have identified eight of the most common equipment and facility design flaws that inhibit the team's ability to master tasks with confidence and efficiency. To avoid potential pitfalls, the following list will help you evaluate your work environment and perhaps encourage you to implement an improvement plan.

Design flaw 1:
Delivery system is inaccessible to both team members.

Rationale
Historically, manufacturers have designed delivery systems around the doctor's needs, rather than for a team approach. This is not to suggest research and

design neglect, but rather to illustrate the ramifications of meeting the customer's needs. Much design is driven by the purchaser—namely, the doctor. In other words, the doctor is the one typically negotiating with the dealer to satisfy his or her needs as an operator, rather than the needs of the "team."

To that end, side-delivery systems continue to be sold, even though the chairside assistant cannot access it. Likewise, over-the-patient delivery causes similar problems with assistant access and is quite cumbersome (dangling over the patient). Both designs also force the operator to leave the work field, thereby causing wasted motion and unnecessary eye fatigue, particularly when using magnification.

Recommendation
With the goal of delivering the highest quality of care—in less time with less stress—it is suggested that the delivery system be positioned closer to the work field and within the chairside assistant's reach, whenever possible.

This can be accomplished by either repositioning the extension arm closer to the operating field or, if at all possible, remounting the unit closer to the assistant.

Improving unit access will allow consistent four-handed transfers and encourage handpiece and burr changes. This arrangement allows the assistant to be more accountable for both rotary and non-rotary instrumentation. When the assistant is actively involved throughout the procedure, attention span increases, thus encouraging active anticipation and treatment facilitation.

Design flaw 2:
Auxiliary support equipment is mounted behind the assistant's work position.

Rationale
The assistant is primarily responsible for maintaining a clear field of view, so support equipment must be easily accessible at all times. Considering that the high-volume evacuator and air-water syringe can be active throughout a procedure, positioning and access become critical.

Units that have been designed with vital armamentarium hanging behind the user create a huge disadvantage of lost time, wasted motion, and added stress for the entire dental team.

An assistant who has to retrieve and replace support equipment from behind will exacerbate musculoskeletal discomfort by twisting and turning

Clinical Equipment

to the rear—or by using his/her lap as a "holding station." This temporary lap position also becomes a greater problem in regard to maintaining a sterile field.

All in all, dental units that place the assistant's equipment behind the assistant can cause lost time, wasted motion, and demand extra time in asepsis.

Recommendation

Ideally, the high velocity suction and air-water syringe should be positioned directly in front of the assistant. It is extremely important to ensure that the front surface of the cart provides clearance for the assistant's legs underneath. This means that support attachments (e.g., suction, air-water syringe, and handpieces) must be mounted on their respective sides to avoid blocking the front pathway (Fig. 8-1).

In addition, the unit should extend over the assistant's lap, bringing these support items and other instrumentation within reach. Be sure the cart top is adjustable to allow adequate leg clearance for the assistant's legs; otherwise, poor posture will be exercised to overcompensate for this discrepancy.

Design flaw 3:
Dental units with small or inaccessible work surface space.

Rationale

When the assistant is supplied with a small work surface, the ability to prepare in advance becomes very limited. We have found that inadequate work surface space increases the chance of incomplete set-ups. Inadequate space causes set-ups to be laid out in phases, rather than at the beginning of the procedure, causing lost time with each phase transition.

Furthermore, the assistant is often tempted to overload a small work surface, creating a hazardous and top-heavy work environment. Aside from the danger of accidents occurring, it becomes much more difficult to locate items when anticipated. Considering that timely passing of anticipated armamentarium is essential for chairside efficiency, every effort to optimize this process must be made.

Accessibility of the unit is also important. If the assistant must extend his/her reach to the unit, risks increase for musculoskeletal stress

ALL *the* RIGHT *MOVES*

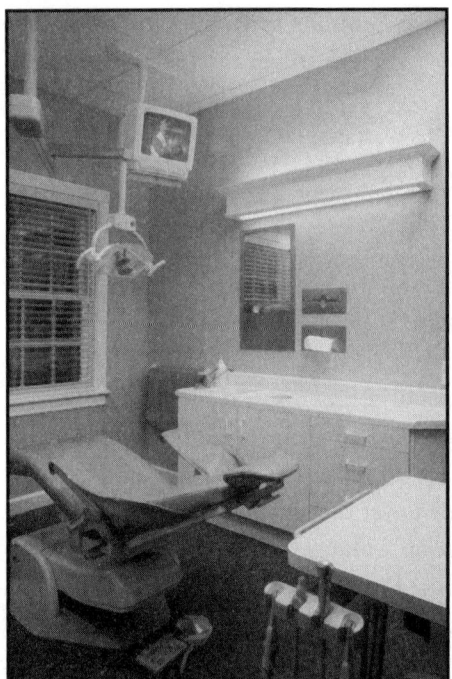

Fig. 8-1 Support attachments (such as the suction, air-water syringe, and handpieces) must be mounted on their respective sides to avoid blocking the front pathway. *Courtesy of Dr. Michael Unthank, Architect*

and strain. Furthermore, these inefficient work habits contribute to lost time and undesirable performance levels—even overall job burnout.

Recommendation
The ideal work surface provides maneuverability, adapts to the worker, and provides adequate storage space. The countertop size should allow instrumentation and material management with ease. An ideal size is 21 by 24 inches. Side cabinetry should also support the preparation and mixing and storage of support materials (Fig. 8-2).

When using fixed cabinetry, consider adapting a pull-out arm that extends over the assistant's lap. If you are currently using a mobile cabinet, consider using the sliding top in the open position to allow for knee space underneath. When using mobile cabinets with the well in the open position, remove or cover items stored in the well to prevent cross-contamination.

Design flaw 4:
Cuspidors are attached to the patient chair.

Rationale
There is a long-standing practitioner belief that patients would panic if a practice did not provide chairside cuspidors. This is an over-exaggerated

Clinical Equipment

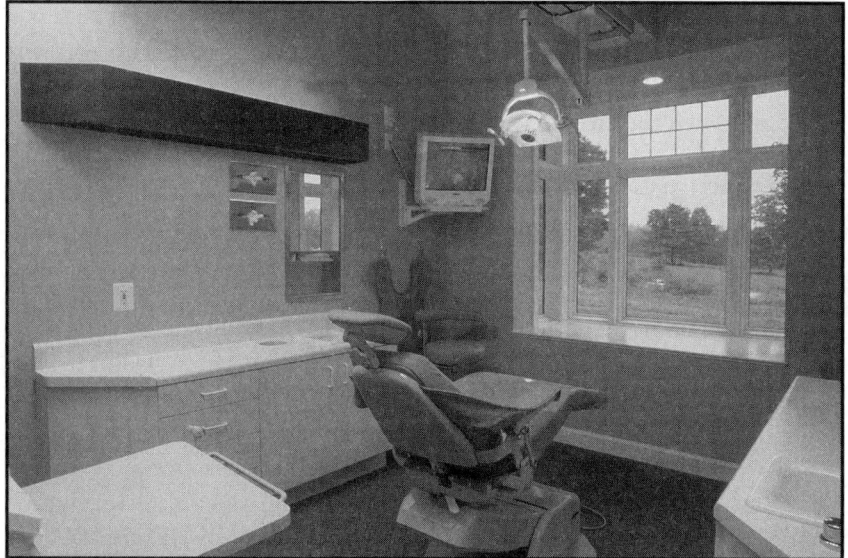

Fig. 8-2 Side cabinetry should also support the preparation and mixing and storage of support materials. *Courtesy of Dr. Michael Unthank, Architect*

fear. Cuspidors no longer serve a purpose in an infection-controlled environment producing efficient, high quality care.

A cuspidor offers nothing more than a "self-serve" approach to oral evacuation. Every time the patient feels the need to expectorate, the procedure is halted at the patient's discretion—not yours.

Unfortunately, when the patient controls the number of breaks, production on an hourly basis is significantly affected. Additionally, asepsis procedures are challenged when anesthetized patients are required to void their own oral fluids.

Recommendation

Dismount the cuspidor and re-educate your patients to appreciate and depend on the benefits of the high-velocity evacuation system. Patient compliance can be more easily gained when infection control benefits are reiterated.

For added patient compliance and neck support, we suggest that you provide the patient a neck pillow, along with a saliva ejector. This will help them feel more in control. The patient can signal the team by raising the saliva ejector when they feel they are in need of supplemental evacuation.

Design flaw 5:
Abdominal support-designed assistant's chair.

Rationale
The abdominal support arm (also known as the belly bar) appears to be the only method of support preventing the assistant from falling out of his/her chair when leaning forward. However, we have found abdominal support arms to actually cause more harm than good.

For example, it is believed that extended use of an abdominal rest can put excessive pressure on abdominal organs. Abdominal rests also encourage the user to sit to one side, or to slouch. This unbalanced posture also puts extra pressure on spinal discs by deviating from neutral sitting postures. As a result, a myriad of associated physical problems can be manifested from abdominal support arms, and therefore should not be used.

Typically an assistant requests an abdominal support arm because he/she is struggling to see around obstructions that block vision. As you can imagine, an assistant working without the bar could very easily fall out of his/her chair while leaning forward. It is important to realize that the bar is only giving the *illusion* of security, and not eliminating the cause of the problem, which is *poor visual access to the operating field*. Hence, the need to improve visual access is what is called for here.

Recommendation
To eliminate the need to lean forward, eliminate the factors blocking the assistant's view.

This goal typically requires a simple re-adjustment of the doctor's mirror-hand fingers. If the doctor simply repositions his/her fingers on the mirror hand—placing them further back on the mirror handle (opposite the mirror head)—access and visibility can be dramatically improved. Additionally, when the assistant is provided an adjustable chair with a wide base, lumbar support, and a foot ring (and is positioned with his/her left hip parallel to the patient's left shoulder), he/she can sit back more comfortably in a balanced position.

Clinical Equipment

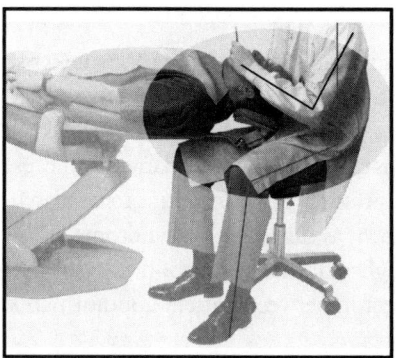 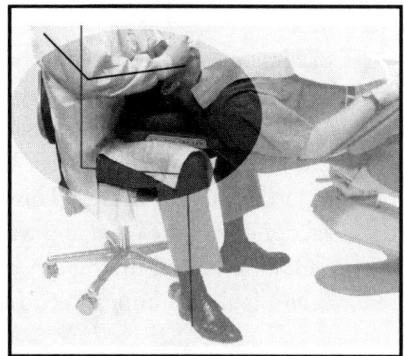

Fig. 8-3 Articulating headrest impinges on the operator's legs. *Courtesy of CRA*

Fig. 8-4 Thin, narrow chairback design allows elbows to maintain 90-degree angle. *Courtesy of CRA*

Design flaw 6:
Thick chair backs force a patient's head higher than the operator's arms, making it difficult to access the operating field without raising elbows. In addition, protrusions from articulating headrests impinge on the operator's legs (Fig. 8-3).

Rationale
Patient chairs have been designed with wider, thicker backrest cushions and protruding control knobs for articulating headrests. These oversized features (which often "sell" patient comfort) ignore important ergonomic needs of the operating team.

Thicker chair backs with protruding controls require that the patient chair be raised to over-compensate for the extra clearance required, and place the operator's hands above elbow height.

Recommendation
Select patient chair designs that offer thin/narrow chair backs (Fig. 8-4). Replace articulating headrests with non-articulating headrests, or use a neck pillow to help position hands at elbow height.

Design flaw 7:
Dental units with coiled handpiece tubing.

Rationale

Coiled tubing can greatly contribute to stress and strain on the hand and wrist. The stress is caused from the memory in the coil, which tugs and pulls on the operator's hand. This can cause the operator to increase pinch forces, over-flex, or extend the wrist to maintain a grasp. Tendons can also become inflamed and cause pressure to the median nerve. This extra pressure can contribute to a musculoskeletal disorder of the hand and wrist known as carpal tunnel syndrome.

In addition, coiled tubing can be a nightmare with infection-control procedures, as it is almost impossible to disinfect in between patient appointments.

Recommendation

Replace all coiled tubing with straight, smooth outer surface tubing fabricated with one continuous seamless outer sheath.

Design flaw 8:

Chair controls are not accessible for dual operation. Rheostat cords can be inaccessible or interfere with stool casters and foot positions.

Rationale

Whether the patient chair is operated by foot or chair control, the controls must be accessible to both the chairside assistant and the doctor.

Most foot controls extend out from under the chair on the operator's side, making this task almost impossible to be performed by the assistant. The rheostat cord can also play interference with stool casters, unit casters, and individual foot positions.

Recommendation

Select chair designs with controls that can be operated from either side of the chair. Choose configurations that center the output at the 12 o'clock position, directly behind the patient chair, or provide dual controls on both sides. Run the rheostat cord under the floor slab so that access begins at the midline of the base of the patient's chair (Fig. 8-5). This will prevent stool caster interference and floor path obstructions while improving access to foot controls.

Other enhancements

In addition to standard treatment room equipment and proper positioning, there is one additional support item that will dramatically enhance the overall benefits of ergonomics. This support item is known as magnification.

Clinical Equipment

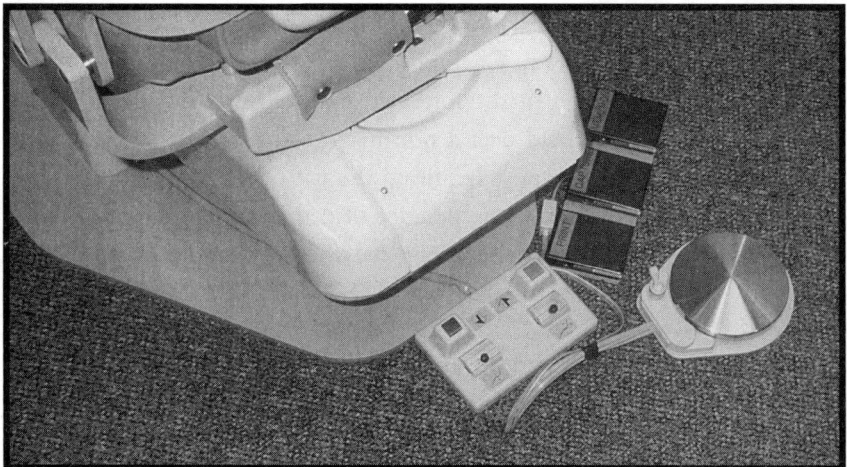

Fig. 8-5 Rheostat cord is run under slab floor to provide access at the midline of the base of the patient's chair to prevent stool caster interference and path obstructions. *Courtesy of Dr. Michael Unthank, Architect*

Historically, practitioners of an older generation had shown interest in field magnification; however, the optics back then (known as "loupes") were not of the quality now seen with second- and third-generation designs.

The concept of magnifying the field was initially "viewed" as a means to improve degenerative eyesight, rather than a quality enhancement tool.

Now that magnification manufacturers offer customizable working distances—convergence angles, viewing angles, power levels, and illumination options—a more sophisticated approach can be adopted. These features now improve working postures and reduce stress factors created from the ongoing struggle encountered by individuals desiring a closer view.

Magnification can put the operator in what appears to be an extremely close range of view without changing posture (flexing forward). Furthermore, attention to detail and overall quality are dramatically improved with a magnified field.

It should be noted, however, that improperly selected magnification systems could support, or even create, poor working postures. Therefore it is critically important that the optics are properly designed and fitted to the individual operator. (Refer to Chapter 5)

If given the choice, a quality-oriented clinician would indeed elect to operate in a magnified field over a non-magnified field. To that end, progressive dental schools are now requiring the use of magnification early on in the curriculum, similar to the practice of using indirect vision and preclinical gloves to better prepare students for direct patient care.

Workspace

Design of equipment and the arrangement of workspaces are key factors in achieving efficient and safe work practice. Equipment should be designed for range of use within ergonomic work spaces.

To that end, individuals designing equipment and workspaces must be aware of ergonomic goals and regulatory requirements. The designer must also match the various parts of the equipment with the function and the abilities of the individuals who perform tasks with the equipment.

It is equally important to include the entire team in the design and selection process to ensure that all individuals' needs have been addressed before equipment is purchased or installed.

STEP NINE

Administrative Workspace

"Until input (thought) is linked to a goal (purpose) there can be no intelligent accomplishment."

— Paul G. Thomas

Workstation design

Many administrative jobs are characterized by long hours of work in a seated position in front of visual display terminals (VDT). Common health complaints associated with working with VDTs include eye strain and eye irritation. Eye problems can be a result of poor lighting, uncomfortable temperatures, air flow, humidity levels, and air quality. Employers can improve the health and comfort of those working with VDTs by properly designing workstation equipment, supporting seated postures, by reducing repetitive hand motions.

The keyboard

The keyboard height should be comfortable for the worker. When it is at an appropriate height, the operator's upper and lower arms form a comfortable 90-degree angle at the elbow. It has been recommended that the keyboard not be thicker than 2 inches at the home row. If the keyboard is too thick, wrist rests may help. It should be noted however, that wrist rests can also exacerbate inflamed areas of the wrists by placing unnecessary pressure on that area. Ideally, keyboards should be positioned in a keyboard drawer (slightly below counter height) to enable the operator to maintain a 90-degree angle at the elbows and neutral wrist positions (Fig. 9-1).

ALL *the* RIGHT *MOVES*

Fig. 9-1 Keyboard is accessed from a keyboard drawer (slightly below counter height) to enable the operator to maintain a 90-degree angle at the elbows and neutral wrist postures. *Courtesy of Dr. Michael Unthank, Architect*

The mouse

The mouse should be on the same plane as the keyboard; therefore, the keyboard drawer must be wide enough to accommodate both the keyboard and the mouse. As a word of caution for those attempting to save space with integrated "rollerball" or "touchpad" keyboard designs—static mouse positions can put excessive pressure on the palm and cause hand fatigue.

The screen

Reflection and glare from the PC screen are frequent causes of eyestrain for VDT users. Some of the most common sources of light interference and glare come from windows and bright, indoor, overhead lights.

Ideally, the user is positioned perpendicular to an uncovered window. Low lighting, in addition to anti-reflection filters, screens, and tilt adjustments, can also help to reduce reflections. (Note: Glare is far worse when the user is either facing a window or sitting with his/her back to a window. This is why glare reflection screens are suggested.)

The monitor should be positioned at a distance that allows the user to see without squinting or leaning forward. The angle of the screen should be tilted approximately 5 to 15 degrees, so that the operator can view the screen slightly below the horizontal plane of view. This will prevent the eyes from drying out and place less strain on the eyes while viewing the monitor. Document holders (positioned at the same height as the monitor) will also reduce strain and tension on the neck while increasing productivity.

The chair

Chair design may be the most important element in achieving balanced posture. The chair must be easily adjustable (without tools) to conform to the individual worker's musculoskeletal needs. Adjustments should include seat height and angle, backrest height and angle, backrest and seat tilt tensions, arm rest height and angle, and natural inward curve of the lumbar area.

The chair should provide five mobile casters to provide balance, as well as ease of movement around the workstation.

The seat height should also be adjustable so that the worker's thighs, shins, and feet form right angles at the knees and ankles. If the chair is too high, circulation to the upper thighs will be affected. When this is the case, a footrest should be provided. Ideally, the footrest should slope upward from front to the rear.

Desktop

The desktop height should be set at approximately 30 inches from the floor, with the screen angled at approximately 5-15 degrees to allow the operator to view the screen slightly below the horizontal plane of view. The operator's seat pan should be no more than 20 inches from the floor to accommodate knee clearance under the desktop. Telephones, printers, calculators, and other support equipment should be positioned within comfortable reaching distance.

ALL *the* RIGHT *MOVES*

Fig. 9-2 Telephone headsets are used to avoid unnecessary tension in the neck and shoulder – and increase productivity.

Telephones

When using telephone handset-style receivers, hold the receiver parallel to your ear with your hand. Do not use your shoulder to balance the handset up to your ear. Ideally, telephone handset-style receivers should be replaced with headset or earset-style components. These designs avoid unnecessary tension in the neck and shoulders—and increase productivity (Fig. 9-2).

STEP TEN

The Master Design Plan

"If you don't know where you are going, how can you expect to get there?"

— Basil S. Walsh

Seamless integration of ergonomic factors within a clinical support center increases efficiency and ensures safety on the job. When these components are not prioritized in the grand scheme of dental office design, performance levels and musculoskeletal health suffer greatly.

Ideally, the dental environment should be designed based on health, safety, and human engineering. It is equally important that support equipment and instrument management systems be linked into the overall design equation to achieve optimum team performance. Team performance therefore is directly related to ergonomic applications within workspace environments.

History

The study of ergonomics dates back to World War II, during which time ergonomic principles were applied to eliminate problems from the operation of complex military equipment. The benefits were so impressive that ergonomic concepts quickly expanded into business environments throughout Europe and the United States.

It was not until the early 1940s that dentistry became aware of the benefits of ergonomics and consequently applied the rationale to auxiliary utilization. At that time, it was believed that a trained auxiliary could increase productivity and minimize musculoskeletal stress for the dentist. This was accomplished by encouraging the practice of four-handed, sit-down dentistry rather than two-handed, stand-up dentistry.

This new mode of operation indeed proved to be more efficient and less fatiguing back then, and continues to be the case. However, even though ergonomic interventions reduce stress and strain, the National Institute for Occupational Safety and Health (NIOSH) continues to record musculoskeletal and nervous system disorders in workplaces across America—dentistry included. It is believed that these disorders are partially ascribed to poor design of workspace and equipment, thereby creating a call for ergonomic applications in dental office design.

More recently, the study of ergonomics caught the eye of the Occupational Safety and Health Administration (OSHA) in regard to the reported increase in work-related repetitive motion injuries; hence, ergonomic guidelines were developed. These guidelines address concerns about repetitive motion activities and poor postures that cause or aggravate symptoms of pain or discomfort commonly found in the neck, shoulders, hands, wrists, and lower back area.

To minimize these risks, employers are encouraged to work with manufacturers and design consultants who specialize in dental office design to help develop workstation improvement programs.

Treatment room environment

The primary focus in the dental office should begin with the core production center, known as the treatment room environment. Each treatment room environment must seamlessly integrate technology, ergonomics, asepsis, and instrument and handpiece management systems into the master plan.

The old adage "form follows function" must become the guiding light to ensure optimal performance and secure operating team safety within each workspace. All treatment rooms should be standardized to maintain consistent procedures. Rooms should be designed to facilitate safe and efficient four-handed and six-handed sit-down dentistry. All instrumentation should be positioned within arm's reach of the dental assistant.

The Master Design Plan

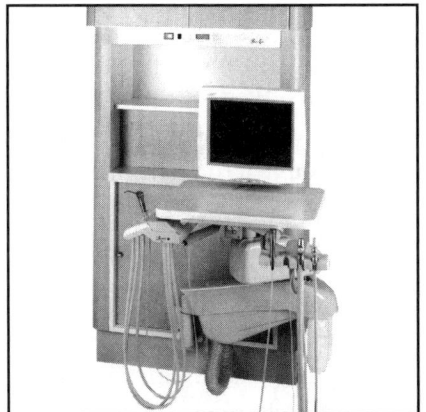

Fig. 10-1 An optimal dental unit operates from a single-use, dual-function, rear-delivery cart to provide the assistant an ideal work surface while maintaining operator access and minimizing contamination otherwise seen in a split-cart system. The single-use, dual-function system also offers a substantial savings in equipment and plumbing when compared to a split-cart system. *Courtesy of Pelton Crane/DCI*

The dental unit is optimized when operating from a single-use, dual-function, rear-delivery cart. This system provides the assistant an ideal work surface while maintaining operator access and minimizing contamination otherwise seen in a split cart system.

The single-use, dual-function system (Fig. 10-1) also offers a substantial savings in equipment and plumbing when compared to a split-cart system.

Automation

Computer monitors and keyboards that support practice management software should be positioned directly behind the patient for optimum access.

Practice management monitors and keyboards. Hardware should be mounted in such a way to provide flexibility in positioning for both the doctor/hygienist and the assistant (Fig. 10-2). They must be thin and compact.

Extension arms. Flexibility is key here. Extension arms that support monitors and keyboards should telescope, pivot, and tilt.

Cordless keyboards, touch-screen monitors, light-pen operation, and voice activation. These features and devices further enhance the efficiency of data input by minimizing the extent of reach and use of the keyboard, saving time and motion.

Patient viewing monitors. These are used for patient education and case presentation, and are best mounted on the wall or ceiling nearest the toe of the patient's chair. This will enable the patient and the operator to view images at the same time, with the patient either upright or fully reclined.

X-ray units. These units are ideally positioned so that the x-ray head can be manipulated from either side of the chair. The x-ray control must also be positioned to provide the operator protection from radiation during exposure. The location should be evaluated prior to mounting to ensure that the cone or

ALL the RIGHT MOVES

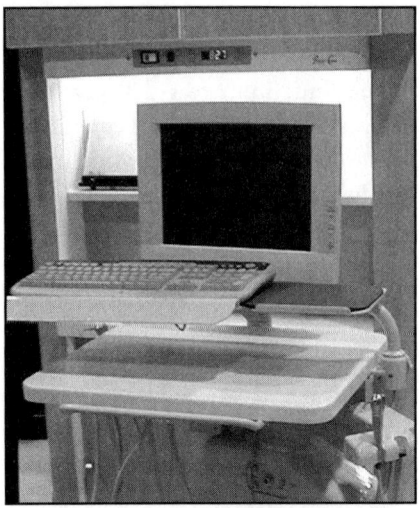

Fig. 10-2 Monitors and keyboards should be mounted on an extension arm to provide flexibility in positioning for both the doctor/hygienist and the assistant. *Courtesy of Pelton Crane/DCI*

extension arm does not interfere with other items, such as computer monitors, keyboards, towel dispensers, soffits, counter-based, or wall-mounted equipment.

Cabinetry

Cabinetry should be constructed of smooth, non-porous materials, with minimal seams for ease of cleaning. To meet this objective, an integral covered backsplash must be specified.

Wood trim and heavily textured surfaces should be avoided to inhibit the collection of microbes. Manmade solid-surface laminates have become the materials of choice. To achieve a seamless result, a process known as "liquid seaming" must be specified.

It has been noted that light-colored and light-textured materials can be more forgiving against scratches than smooth, dark, or high-gloss surfaces. Dark-colored surfaces also tend to show more dust and fingerprints. Be mindful of minimizing laminate patterns, as they may pose a problem when attempting to locate miniature support items, such as burrs on the countertop.

Low-pressure Melamine® or LPM is considered to be an excellent choice for the inside of cabinet drawers and shelving, due to its easy cleaning properties, which are created in manufacturing.

LPM has a paper face that is saturated in a liquid plastic and "thermally fused" under heat and pressure. This fusion into the substrate creates a molecular bond, and the paper becomes impregnated with the panel that enhances its abrasion resistance. LPM is preferred to softer

The Master Design Plan

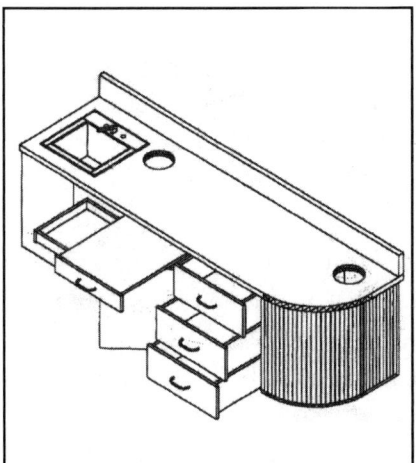

Fig. 10-3 Cabinetry must be designed and strategically installed with human engineering factors kept in mind and applied to the treatment environment. The height, depth, and length, along with the exact position of the cabinetry, number of drawers, depth and clearance inside each drawer, and so forth, must be coordinated to ease access to countertops and provide adequate drawer storage space. *Courtesy of Sullivan Schein Design*

materials, such as vinyl, as it appears to be more durable and holds up better when exposed to scratches or chemicals.

Cabinetry must be designed and strategically installed with human engineering factors kept in mind and applied to the treatment environment. The height, depth, and length, along with the exact position of the cabinetry, number of drawers, depth and clearance inside each drawer, and so forth, must be coordinated to ease access to countertops and provide adequate drawer storage space (Fig. 10-3).

It should be noted that the number of drawers and their contents should be minimized to simplify infection control requirements. Streamlining storage areas can also reduce the temptation to access drawers during procedures and minimize cross-contamination.

Operational pulls also play an important part in the function of cabinetry with respect to infection control. Spring-type "touch latches" and standard pulls are options that should be considered.

While touch latches can offer ease of operation with their "touch-anywhere-to-open" feature, they tend to challenge decontamination procedures by expanding potential areas of contamination.

Drawer and cabinet pulls keep potential contamination to the pull itself. Therefore, it is suggested that bar-type pulls be used.

Bar-type pulls also allow individuals to use an auxiliary forceps as a means to operate the drawer. A knob-type pull typically requires an entire

ALL *the* RIGHT *MOVES*

hand to manipulate the drawer.

Full extension glides and hinges improve access to items in the back of drawers. This can maximize the use of cabinet space and minimize risk of injury from "blind" retrievals. It is also important that hands-free counter top waste drops be provided to minimize countertop clutter and ease the process of disposal (Fig. 10-4).

Treatment room sinks should be designated solely for use by operating team personnel for hand washing and material preparation. Treatment room sinks should not be used for patient rinsing to control asepsis.

Ideally, sinks should be operated with minimal handling—as nearly a hands-

Fig. 10-4 It is important that hands-free counter top waste drops be provided to minimize countertop clutter and ease the process of disposal. *Courtesy of Dr. Michael Unthank, Architect*

The Master Design Plan

free process as possible. This would include foot controls or motion sensors and/or a single-lever faucet control. The single-lever faucet control can be used to activate the flow, adjust the temperature, or as a backup in case of sensor malfunction.

Note: Motion sensors are known to fail and quite often do not recognize dark colors.

If using solid-surface countertops, an integral solid-surface bowl may be used. Otherwise, a high-quality stainless steel basin (with high nickel content to minimize rusting) should be used.

Dispensers

Wall-mounted dispensers help minimize countertop clutter and improve ease of access during use. For example, recessed soap dispensers can be mounted to the sink to make hand-washing procedures easier and to minimize clutter around sink areas.

In addition, gloves, paper towels, and cup dispensers help maintain countertop cleanliness and minimize cross-contamination generated from aerosols during treatment.

Dispensers can be made of the same laminate-type material as the corresponding countertop for surface mounting, or recessed into the wall above the base cabinet in stainless steel housings.

Floor coverings

Floor coverings made of a hard surface, such as seamless linoleum, offer optimum cleaning and disinfection functionality. However, it should be noted that hard-surface flooring can be less secure for the operating team when seated on stools with mobile casters. This is of particular importance when the floor is not level.

Furthermore, hard-surface flooring will not absorb noise as well as carpet, nor will it offer the warmth that a carpet floor covering provides. Currently, OSHA does not regulate floor covering, and so the selection of materials continues to be at the business owner's discretion. When carpet is used, a high-gauge, low "face weight" 6,6 nylon yarn fiber is recommended. Nylon carpet is more durable than other yarns, such as olefin or polyester.

Type 6,6 nylon offers a harder and denser molecular structure, making it more soil resistant and better able to maintain resiliency. Carpet yarn may be yarn-dyed or solution-dyed. The selection is often determined by aesthetics.

Yarn-dyed carpet provides a wide range of vivid colors; however, it can lose coloration from chemicals such as bleach. Solution-dyed carpet is more resistant to bleach. It should be noted, however, that while bleach may not color fade solution-dyed carpet, it can deteriorate the fiber.

It is also important that the carpet offers—

- a high number of stitches per square inch (SPI)
- be of a low pile
- provide a closed-cell vinyl backing

The carpet should specify a closed-cell, moisture-proof, integral bonding process with vinyl backing.

The carpet should be glued directly to the concrete slab floor. Pads should not used, as they can attract and retain moisture, which can act like a sponge upon impact, further promoting mildew and microbial contamination.

Historically, wall coverings have been made of upholstered materials that have been found to be dimensionally unstable and difficult to clean. While they may offer more acoustic absorption, they can be less durable and are more apt to tear or rip than vinyl wall coverings.

For maximum durability, clinical areas should specify a type-2 vinyl. Non-patient areas are sufficient with a type-1 vinyl.

Instrument recirculation

The instrument recirculation center should be centrally located to all treatment rooms. The counter space must provide adequate space for both pre-cleaning and sterilization procedures. Engineering controls—such as ultrasonic cleaners and instrument cassettes—provide additional protection by minimizing the need to handle contaminated instruments.

Cassettes come in many different sizes for different procedure needs (Fig. 10-5). Cassette instrument management systems eliminate handling and sorting, which saves time and dramatically reduces exposure risks.

For offices concerned with size and space, compact designs known as "double stacking" or "two-tier" systems double the instrument capacity of standard arrangements.

Sinks and ultrasonics

Large, double-compartment, stainless steel sinks (with high nickel content to minimize corrosion) can provide simultaneous presoaking and rinsing. Hand-scrubbing should be avoided; therefore, ultrasonic units should be utilized to minimize the handling of contaminated, reusable sharps.

Ultrasonic units should be recessed into the countertop to eliminate the need to lift items up and out of the unit. To ease the process of emptying the tank chamber, either plumb the unit directly into the sink drain or plumb a gate-valve from the drain hose of the tank to the central vacuum system.

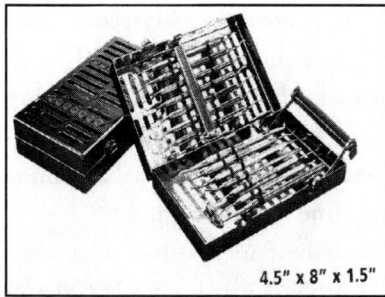

Fig. 10-5 "Two-tier" cassettes double the capacity of instrument storage when using compact designs. Cassette instrument management systems eliminate handling and sorting, which saves time and dramatically reduces exposure risks. *Courtesy of Hu Friedy*

Dr. Mark J. Friedman of Encino, California, initiated the central vacuum system hook-up in his dental office several years ago and found this unique plumbing arrangement substantially expedited unit draining. Dr. Friedman also installed a separate faucet above the ultrasonic unit to expedite the process of refilling the unit (Fig. 10-6).

The internal and external surfaces of all the base and upper cabinetry should be constructed of smooth, nonporous materials, with minimal seams for ease of cleaning and disinfection. Manmade solid-surface laminates have become the material of choice. To achieve a seamless result, "liquid seaming" must be specified.

The countersplash should be continuous from the counter to the upper cabinetry to avoid cleaning painted wall surfaces. Use light-colored surfaces, as dark-colored surfaces are hard to clean and tend to show more dust and fingerprints.

In addition, all interior surfaces should be lined with low-pressure Melamine® to provide more durability to potential scratches or chemical exposures.

ALL *the* RIGHT *MOVES*

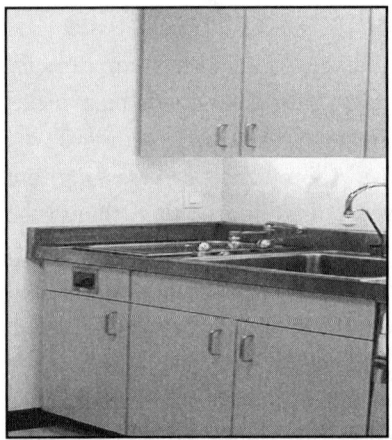

Fig. 10-6 Ultrasonic unit uses a gate-valve (from the drain hose to the tank to the central vacuum) to more efficiently drain the unit. When a primary sink faucet extension is not available, a separate faucet above the ultrasonic unit can expedite the process of refilling the unit. *Courtesy of Dr. Mark J. Friedman, Encino, CA*

Countertop waste receptacles can facilitate waste disposal with minimal handling. Countertop drops reduce potential cross-contamination during disposal and minimize time and motion.

It is best to have three separate waste drops. One receptacle is designated for regular waste, another for regulated (red bag) waste, with the third used as a central "backup" for miscellaneous sharps.

It should be noted that when using countertop drops, all biohazard waste containers and sharps containers must be securely adapted to the countertop grommet and covered in between active use.

Ventilation

Efficient ventilation in the instrument-processing center must also be considered to avoid odors that cause eye irritation, headaches, or potential respiratory problems. Air should be exhausted, while fresh air is returned back into the system. Daily monitoring of air quality function and routine filter checks are recommended to achieve ideal conditions within enclosed work environments. (See "Ventilation and air quality," page 66.)

The lab

Laboratory areas must provide adequate space for grinding equipment, stone storage, impression pouring, model trimming and storage, and vacuum exhaust. Cabinetry should be constructed of smooth, non-porous materials with minimal seams for easy cleaning.

To meet this objective, an integral, covered backsplash must be specified. Wood trim and heavily textured surfaces should be avoided to inhibit the collection of microbes. Manmade, solid-surface laminates have become the materials of choice.

Once again, consider light-colored and light-textured materials, as they can be more forgiving against scratches than smooth, dark, or high-gloss surfaces. Remember, dark-colored surfaces also tend to show more dust and fingerprint.

Interior surfaces should be lined with low-pressure Melamine® to support cleaning requirements. The countersplash should be continuous from the counter to the upper cabinetry to minimize difficulty in cleaning painted wall surfaces.

Floor coverings in the lab should be a commercial-grade hard surface. Seamless linoleum (vinyl) offers optimum cleaning ability in the lab, while rubber offers an additional benefit of being slip resistant.

It should be noted that hard-surface flooring does not absorb noise as well as carpet. It should also be noted that a hard-surface floor can actually transfer dirt and dust more easily to carpeted areas.

Ventilation efficiency in the lab must also be considered to avoid odors that cause eye irritation, headaches, or potential respiratory problems. Air should be exhausted while returning fresh air. Vacuum systems should be provided to capture dust particles associated with general grinding and trimming tasks.

The darkroom

The darkroom must include an entrance with a light-proof seal. Adequate space must be allotted on the countertop for an automatic processor and film duplication devices. A deep sink with a water supply for cleaning rollers, along with a second overflow drain and approved capture containers, should be provided to avoid water damage and ease the process of capturing solutions for "off-site" transfer and treatment.

The darkroom must also provide adequate ventilation to ensure that chemicals used within this space do not cause harm to individuals. Floor coverings in the darkroom should also be a commercial grade, hard surface such as seamless vinyl or rubber to provide easy cleanup for chemical spills.

State-of-the-art dental offices using either daylight loaders or digital radiography have virtually eliminated the need for a dark room.

Public traffic areas

Traffic flow must prohibit patient passage through laboratory workstations, instrument processing, x-ray processing areas, or other treatment rooms. Patients must directly access treatment rooms from the reception area.

The corridor width must comply with building codes and provide for passage of patients and healthcare workers without injury or danger of cross-infection.

In addition, a "holding area" coat rack should be provided in an alcove of the main hallway that leads to non-clinical areas. The objective of this holding area is to provide a quick "hang-up" area for soiled garments, as required by OSHA when entering non-clinical areas of the dental office.

The staff lounge should be used as a break room or lunch area only. More recently—due to laundering requirements set by OSHA—staff lounges have expanded into laundry and changing areas. Therefore, it is not uncommon to find a washer and dryer, storage, and uniform closets in this area.

Restrooms (staff and doctor)

The staff and doctor restrooms allow clinical personnel privacy, away from public facilities. These restrooms are best located within the clinical section of office to minimize travel.

The staff restroom can also provide an additional changing area and, if necessary, storage for clean garments.

The Master Design Plan

Laundry

Uniforms can be laundered on-site with minimal handling. In-office washer/dryer units and hampers should ideally be stored away from eating areas or patient-care activities to prevent cross-contamination.

When space does not allow the separation of laundry activities from other activities, laundry must be managed off-site unless clear delineation can be identified to further prevent contamination. To help meet this objective, each office should have a dedicated hamper, which is properly labeled, and a dedicated closet for the storage of clean garments.

Fig. 10-7 Mechanical rooms house the compressor and vacuum. They should be strategically located to reduce sound transfer to patient areas. *Courtesy of Dr. Mark J. Friedman, Encino, CA*

Mechanical equipment rooms

Mechanical rooms (housing the compressor and vacuum) should be strategically located to reduce sound transfer to patient areas (Fig. 10-7).

Current thinking advises against storing both the compressor and vacuum in the same room, as the exhaust of the central vacuum can contaminate the compressed air.

Oil-bearing compressors should be drained weekly, and filters must be changed periodically to further minimize cross-contamination.

This room must be equipped with appropriate ventilation. It is also important that mechanicals—such as the water heater, telephone equipment, and electrical services—be stored separately from the compressor and vacuum in their own utility location.

Nitrous oxide storage

Nitrous oxide cylinders must be stored away from potential ignition sources and be secured (individually chained to a fixed surface) to prevent tipping. Nitrous oxide cylinders must be stored in an area separate from all other mechanical storage rooms, and must prohibit pass-through to other areas.

ALL the RIGHT MOVES

Nitrous oxide storage rooms must also provide a fire-rated door and lock set with appropriate ventilation in the door, as specified by the National Fire Prevention Association (NFPA). When nitrous oxide is stored remotely, it must also have a zone "cut-off" valve.

Ventilation and air quality

Heating, ventilation, and air conditioning (HVAC) systems are designed to provide air at comfortable temperature and humidity levels, free of harmful concentrations of air pollutants. Of these three elements, the ventilation system continues to be the most complex process involved in determining the quality of indoor air.

Ventilation is a combination of processes through which the supply and removal of air from inside a building occur. During this process:

- air is brought in from the outside
- the air is conditioned and mixed with a portion of the indoor air
- the "mixed" air is distributed throughout the building, while a portion of the inside air is exhausted outside

The quality of indoor air may deteriorate when one or more of these systems functions inadequately. An example of a poor operating system would be one that allows an accumulation of carbon dioxide (which is produced when people breathe). Carbon dioxide is a pollutant that may cause building occupants to grow drowsy, suffer headaches, or function at low activity levels.

So-called indoor air pollution is primarily caused by an accumulation of indoor contaminants that are not properly exhausted. Common pollution sources include tobacco smoke, biological organisms, building materials and furnishings, cleaning agents, dust from copy machines, and pesticides.

HVAC systems that are improperly maintained can contribute to serious conditions such as Legionnaire's disease and sick building syndrome (SBS).

While Legionnaire's disease has identifiable causes, SBS can display physical symptoms without clearly identifiable causes. SBS symptoms may include dry mucous membranes and eye, nose, and throat irritation, all of which can lead to reduced work efficiency.

The Master Design Plan

Historically, building occupants would control potential pollutants by opening windows to air out stuffy rooms. Today, most office buildings are constructed without operable windows and with mechanical ventilation systems used to exchange indoor air with a supply of relatively cleaner outdoor air.

Outdoor air rates are supplied into a building according to the city's building code, which is primarily based on the need to control odors and carbon dioxide. Carbon dioxide is essentially a component of outdoor air, but its excessive accumulation indoors can indicate inadequate ventilation.

The American Society of Heating, Refrigerating and Air Conditioning Engineers (ASHRAE) published its Standard 62-1989 to address this issue. "Ventilation for Acceptable Indoor Air Quality" is a voluntary standard intended to avoid adverse health effects from indoor air. The standard applies to all types of facilities, from dry cleaners to hotels, as well as convalescent hospitals.

The standard specified rates at which outdoor air must be supplied to each room within the facility range from 15 to 60 cfm/person (depending on the type of activities performed within each facility). When HVAC systems are not properly maintained to promote indoor air quality, ventilation systems can become a source of contamination or become clogged and reduce or even eliminate airflow.

Humidification and dehumidification systems must be kept clean to prevent the growth of harmful bacteria and fungi. Failure to properly treat the water in cooling towers (to prevent growth of organisms such as legionnella) may introduce such organisms into the HVAC supply ducts and cause serious health problems.

Although air cleaners may be an important part of an HVAC system, they cannot adequately remove all of the pollutants typically found in indoor air.

Air cleaners that have a high filter efficiency and are designed to handle large amounts of air are the best choice for use in an office. Air cleaners include simple furnace filters, electronic air cleaners, and ion generators.

Filters are either flat or pleated, and are used for removing particles in the air. Flat filters collect large particles, while pleated filters—such as the high-efficiency particulate air (HEPA) filters—collect smaller particles. Electronic air cleaners and ion generators use an electronic charge to remove airborne particles.

ALL the RIGHT MOVES

Note: These devices might also produce ozone, which is known as a lung irritant. To safely improve air quality, it is recommended that you perform maintenance inspections on a regular basis and maintain records of those inspections, as well as document HVAC system-related problems.

In addition, it is recommended that you control pollution sources by increasing ventilation rates during periods of increased pollution. It would also be prudent to keep updated on ventilation standards and building codes, as well as reexamining energy conservation practices with regard to employee health and productivity.

In summary

Efficiency and productivity goals must take ergonomics and infection control considerations into account to secure functionality and safety. Dental offices must examine their use of space more carefully to reduce musculoskeletal risks and ensure maximum team performance.

"Rightsizing" workspaces has now become the critical linking mechanism in space planning. Information and technology can provide workers with a physical work environment that actively supports their tasks and ensures safety on the job. Optimizing the use of skilled professionals—consultants, designers, and architects who hold expertise in the area of ergonomics and office safety—can assist the doctor in securing a functional layout.

Likewise, involving employees in the planning gives the team a chance to reengineer their workspace and tasks for greater efficiency, while reducing potential safety hazards.

STEP ELEVEN
Integrating Technology

"There is nothing in the horizon itself that limits vision, for the horizon opens on to all that lies beyond itself."

— James Carse

Beyond the horizon

Welcome to the horizon of the 21st century, where you and your team can simultaneously access patient records, digital images, and a plethora of practice management information at the click of a mouse or tap on your screen.

Those who have integrated technology successfully have become the envy of those who were once disbelievers. This is not to say that there have not been learning curves and/or glitches that had to be worked out. Nevertheless, these proud pioneers have allowed us to fail *forward*, by providing us with the truth of their journey and the discoveries of their learning experiences.

In spite of these success stories, hundreds of offices continue to operate well below optimum levels with outdated equipment and software. Environments devoid of total integration will continue to play havoc with efficiency and productivity.

The beauty of total integration is that it allows "once entered" treatment to become the nucleus to all subsequent procedures, such as scheduling, treatment documentation, posting, billing, and reporting. Likewise, letters, reports, and insurance submissions can be "ordered up" immediately while posting activities that warrant follow up. In addition, electronic patient records can be remotely accessed from satellite offices and outsourced for outcome analysis, as well as e-mailed to specialists for diagnosis, interdisciplinary treatment planning, and educational purposes.

To ensure maximum utilization, we suggest that you engage team members in the evaluation and selection process. This will empower employees and ensure that functional needs are met. Of equal importance, each piece of technology must support operating team ergonomics, practice management efficiency, and the delivery of consistent, high-quality patient care.

The hidden gem gained from a fully networked office is that the office becomes fluent in cross-training, ultimately supporting an atmosphere of teamwork from cooperative, overlapping talent and skill. For the most part, any staff member can handle the majority of all patient-processing procedures, allowing the office to be less vulnerable during vacation or sick leave.

The primary objective is to shift your organizational structure from a multi-station, micro-managed operation to a one-stop, comprehensive care environment.

Decentralizing allows one person to become accountable for handling the patient through the full cycle. This provides the patient with a feeling of continuity, which builds trust and gains long-term loyalty. Likewise the staff can save a significant amount of time when using "single-entry" charting.

Furthermore, single-entry charting seamlessly connects treatment that has been diagnosed with a master treatment plan, financial plan, schedule, and billing. Single-entry automation has been proven to minimize posting errors and to streamline dismissal procedures dramatically.

Integrating Technology

Networks: The Master Design Plan

To digitally network patient profiles and practice management information throughout the entire office, a network connection (such as an RJ-45 network connector) must be made available at each computer workstation. A layout of the office detailing these locations, as well as the hub and the main server location, must be identified during the planning stage to ensure that nothing is left out of the network.

To ensure bandwidth and speed of transmission to and from the server, networking cable (known as Cat 5 *enhanced*) is used. Typically, network lines (for each workstation) are run independently from the main hub up through the ceiling (through the inside of a wall closest to the hub), and then "dropped" down a wall adjacent to each workstation location. It is prudent to secure a schematic of your "cable run" to identify these locations if and when technology needs to be modified.

Note: a certified network installer who understands the intricacies of networking should manage installation and design.

To fully integrate the patient record and optimize patient education and communication, an array of digital devices must be integrated. These devices include, but are not limited to:

- digital intra-oral cameras
- digital radiography and imaging (for enhanced diagnosis and case presentations)
- computerized probes and voice activation for quick and easy aseptic charting

This integration can also include blood pressure monitoring and multimedia patient education (*i.e.,* PowerPoint, DVDs, and interactive CD-ROMs).

ALL *the* RIGHT *MOVES*

Along with the integration of these options, it becomes prudent to ensure that each treatment room's central processing unit (CPU) offers performance speed with plenty of expansion slots for sound and video cards. It is equally important to have accessibility for easy access and connection of external devices to appropriate ports. It should be noted that laptops do not provide this flexibility without great expense and inconvenience.

Developing a standardized scheme for positioning treatment room technologies is critically important to the success of utilization. We have identified four primary zones for this purpose, with combination options.

- *Zone 1* manages the private clinical computer workstation for professional staff access and is therefore positioned behind the patient
- *Zone 2* manages the monitor for patient viewing. It is positioned on the wall or ceiling for patient access in either a reclined or upright position
- *Zone 3* manages all remaining support technology for cameras, x-ray sensors, electronic probes, and blood pressure equipment. These are positioned for easy access of the dental unit, the utility wall, or side cabinetry
- *Zone 4* houses procedure-related items such as handpieces, power scaler, air abrasion, lasers, curing light units, procedure instruments, air-water syringes, and evacuation equipment, typically, part of the dental unit
- *Zone combining*. Manufacturers are combining technology for ease of integration and ergonomic space considerations. For example, curing lights, cameras, and patient viewing monitors can be supplied off the patient chair and / or dental unit.

Zone 1 should ideally be positioned within arms' reach of the operating team to maximize access and visibility and minimize movement. By directly mounting the clinical terminal behind the patient's head (off the back utility wall on a flexible arm that telescopes, tilts, and swivels side to side), the doctor, assistant, or hygienist has the ability to view the screen, manipulate images, or call up information from their seated work positions with minor movement (Fig. 11-1).

Integrating Technology

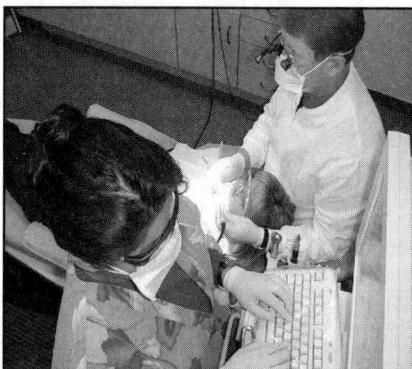

Fig. 11-1 By directly mounting the clinical terminal behind the patient's head (off the back utility wall on a flexible arm that telescopes, tilts, and swivels side to side), the assistant can access the terminal from a seated work position. *Courtesy of Philip Horning, DDS, Chico, CA*

Ideally, the monitor's extension arm should telescope out approximately 12-20 inches (Fig. 11-2); swivel approximately 180° (Fig. 11-3); and tilt approximately 30° (Fig. 11-4). Adding voice activation, touch screen capability (with a sterilizable stylus), and an option for keyboard operation is most beneficial for documenting the clinical record and detailed treatment record notes.

Zone 2 is ideally positioned to provide access for both the patient and operating team when the patient is viewing the monitor in either a reclined or upright position. It avoids patient chair adjustments during co-diagnosis style examinations and presentations. Relying on the upright presentation of a "laundry list" at the end of an examination is not conducive for co-diagnosis, and therefore should not limit presentation options.

It should be noted that each computer monitor needs a separate video card so that a team member could be inputting information at the "team" viewing station while a case is being presented or an educational program is being viewed on the "open" viewing station—simultaneously (Fig. 11-5). If dual video cards are not made available, a source switching device (an AB switch) can be used; however, it will not allow both workstations to function at the same time.

Zone 3 houses all adjunct equipment that will be networked to the main server and becomes part of the electronic chart. Such equipment includes the digital camera, digital x-ray sensors, voice activation devices, BP monitoring equipment, and electronic probes. These items must be positioned within arms' reach of the operating team to ease access. Likewise, the external ports on the CPU must be equally accessible to allow these technologies to "plug and play" without moving cabinetry or surrounding equipment.

ALL the RIGHT MOVES

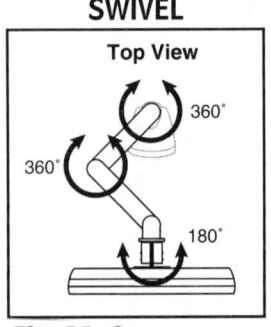

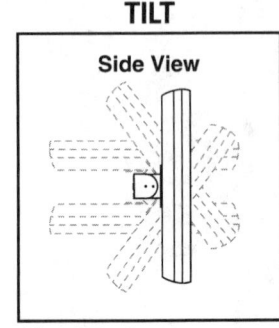

Fig. 11-2 **Fig. 11-3** **Fig. 11-4**

The monitor's extension arm should telescope out approximately 12"-20" **(Fig. 11-2)**; swivel approximately 180° **(Fig. 11-3)**, and tilt approximately 30° **(Fig. 11-4)**. *Courtesy of Ergotron, St. Paul, MN*

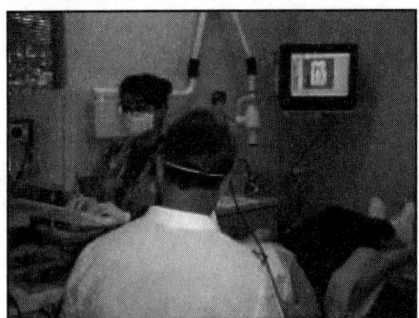
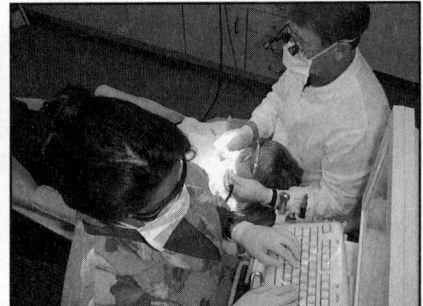

Fig. 11-5 Each computer monitor needs its own video card so that a team member could be inputting information at the "team" viewing station while a case is being presented or an educational program is being viewed on the "open" viewing station—simultaneously. *Courtesy of Dr. Phil Horning, Chico, CA*

Integrating Technology

Zone 4 manages support equipment that is not typically networked to the main server, but rather becomes a part of the dental unit. This includes handpieces, power scaler, laser, air abrasion, curing light, procedure instruments, air-water syringe, and evacuation systems.

Zone combining integrates various components from a combination of active zones. For example, American Dental Technologies combines several technologies from zones 2, 3 and 4. Their microprocessor-controlled electropneumatic base technology offers options of adding all modules at the time of delivery or at a future date—all of which operate from a single foot control (Fig. 11-6).

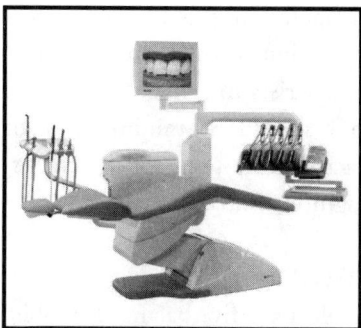

Fig. 11-6 Anthos System® by American Dental Technologies represents a combination of several components from zones 2, 3, and 4, operating from a single foot control. Courtesy of American Dental Technologies

Working smarter

We now know that unbalanced seated postures and the use of intensive upper-extremity movements contribute to a great number of inefficiencies and musculoskeletal health risks. We have seen musculoskeletal disorders as the nation's most serious workplace/occupational health hazard. This number-one rating is primarily due to risks associated with the increased use of input devices and visual display terminals. Therefore, as clinical environments are modified to accept new high-tech treatment modalities, ergonomic factors must be taken into account and employed to enable clinicians to work smarter.

Optimum human function is predicated by the ability not to exceed the structural limitations of one's muscles, tendons, nerves, and joint structures. To that end, each workstation must be analyzed to ensure that these risks are avoided.

ALL *the* RIGHT *MOVES*

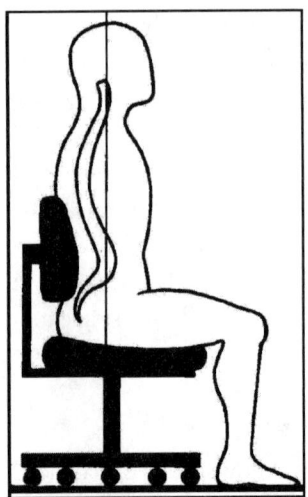

Fig. 11-7 Chairs should have five balanced mobile casters to provide movement and stability. The stools should also provide adequate back support, and the lumbar area should be adjustable in height as well as forward and backward. *Reprinted with permission from Sitting on the Job, Scott Donkin, D.C.*

Visual display comfort

Monitor viewing is less fatiguing on the eyes when monitors are positioned 15 to 32 inches away from the operator's eyes. To prevent glare, the monitor and keyboard should ideally be placed perpendicular to windows and between overhead lights, rather then directly below them.

Keyboard comfort

Keyboard operation should allow a 90-degree elbow position, with forearms parallel to the floor and wrists in a neutral position. Keyboard input should be minimized with the use of macros, auto-text, touch screens, and voice-recognition software.

Chair comfort

Chairs should have five balanced, mobile casters to provide movement and stability. The stools should also provide adequate back support, and the lumbar area should be adjustable in height and depth. (Fig.11-7) The seat pan must be adjustable in both height and tilt for different operator needs. The floor or a footrest should support feet.

Securing your future

Forward-thinking business owners who integrate technology and ergonomics into their master plan will join an exciting, ever-growing group of progressive businesses. Actualizing a fully networked practice will no doubt require an upfront investment of both time and money. Such an investment, however, can be one of the best you'll ever make.

Integrating Technology

With greater accuracy in posting, billing, and tracking—and a huge reduction in the information transfer error factor—the practice experiences a revolving door of benefits. In addition, patients will be rewarded with a feeling of genuine continuity, and the staff is offered a model of partnering for greater accountability through enhanced teamwork.

The foundation of your success will indeed rely upon your willingness to outsource your high-tech needs to experienced professionals and licensed installers.

Your team's ability to embrace technology and the process of cross-training with a sense of privilege and esteemed professional spirit will be predicated upon your ability to cultivate an atmosphere of team input and empowerment.

STEP TWELVE

That's a Stretch!

"My strength lies solely in my tenacity."

— Louis Pasteur

We are blessed with the power and freedom of choice. We control our ability to make decisions to do the things that keep us healthy, active, energetic, and strong. Given that fact, why do most American workers ignore their musculoskeletal health?

Does it take weakness or debilitating pain, or a diagnosis of active disease, to force us to take physical health more seriously? No doubt, physical fitness has become a vital process that helps us sustain our health, strength, and well being. It must not be ignored!

To that end, this chapter has been dedicated to helping workers integrate "fitness" concepts into their daily lifestyles.

Health and Fitness

Fitness is defined as the body's ability to efficiently function. Fitness can be divided into two areas: cardio-respiratory fitness and muscle fitness.

Cardio-respiratory fitness

This is the first, and perhaps most important area of fitness. The heart and blood vessels' ability to transport oxygen from the lungs to the body's tissues determines it. When oxygen is transported efficiently, blood flow can also be transported more efficiently to the organs and systems of the body.

Blood flow is responsible for nutrient supply and the removal of by-products of the muscle metabolism. It has been documented that inefficient removal of by-products from the muscle metabolism can cause tendonitis. Therefore, good cardio-respiratory fitness may help reduce muscle strain while achieving optimum performance.

Muscle fitness

Muscle fitness plays a much different role in the function of the body. Good muscle fitness refers to the muscle's ability to operate at maximum capacity and the overall endurance the muscle can manage.

Muscle fitness can also involve the ability of the muscle to optimize its range of motion. Range and flexibility can be impaired by injury or poor positions that shorten structures by placing tendons in shortened positions and use only limited arcs of motion.

Exercise consultants recommend moderate physical exercise, performed at 80% maximum heart rate for approximately 30 minutes, three to four times per week. Moderate workouts are preferred to vigorous ones, which aggressively engage the muscle groups and thereby exhaust the body.

To complicate matters, as we age, our muscles atrophy. It is believed that after age 35, we lose approximately a half-pound of muscle each year. Along with that muscle loss, we gain a pound and a half of fat each year, all of which contribute to poor physical fitness.

Regular exercise offers improved health and a perception of diminished stress and greater job satisfaction. Physical exercise involves muscle contraction and relaxation, which move the bones at the joints. Movement also allows the joints to remain lubricated for smooth and efficient ranges of motion. Movement also supports spinal health because spinal discs require movement to ensure an exchange of fluids.

Exercise should not be engaged in without adequately and safely stretching, both as a warm-up and finishing as a cool-down. The goal is to stretch and strengthen the muscles and condition your body. The exercise program should boost and sustain your ideal heart rate to help you utilize oxygen more efficiently, improve blood cholesterol balance, lower blood pressure, aid in weight control, and relieve stress. Likewise, exercises should support alignment and should balance both sides of your body, particularly if you tend to favor one side during the workday.

Note: It is advisable to seek professional advice from your doctor before preforming any exercise routine.

Stretching Warm-up

A fit body is flexible, but also strong and well conditioned. Muscles need to be stretched and challenged to remain strong and healthy. However, if any of these movements cause pain or make you lightheaded, seek professional assistance.

A full body warm-up or specific body-part warm-up is advised prior to working out. When a warm-up routine is preformed for approximately 5 to 10 minutes, the deep muscle temperature rises, increasing blood flow, minimizing risks, and increasing flexibility. To that end, it would be prudent to warm-up prior to performing dental procedures. Suggestions for stretching would include the distal forearm, the hand and finger muscles, and the muscles of the shoulder girdle, neck, and back.

Be mindful that over-stretching can cause tears in the soft tissue, which can lead to injury. Stretching should be performed intermittently, three to four times a day; however, it is advisable to seek professional advice from your doctor before engaging in any exercise routine.

The following exercises are examples of areas of focus.

Hand stretches

Warming up the hands can promote the diffusion of synovial fluid, which lubricates the hands and the fingers. This can be accomplished by placing a clock face over the hand. Begin by holding the fingers of the hands towards 12 o'clock. Direct the fingers towards 3 o'clock as you form a right angle at the knuckles.

Next, the fingers are closed into the hand by touching fingers to palm with fingers pointed towards 6 o'clock. After the fingers are closed, form a fist by buckling your fingers tightly into your hand. The fingers are then brought up to the highest portion of the palm. Repeat with the opposite hand. (Fig. 12-1)

Forearm stretches

Begin by holding the arm out at a 45-degree angle to your body. Now, take hold of your right hand with your left hand and flex the right hand back toward your body. Hold this position for 45 seconds (Fig. 12-2). Next, reverse the hand from a flexed position to an extension. With your palm up, press on the fingers of the right hand with your left hand. Hold for another 45 seconds.

Now reverse hands and repeat this exercise.

If you are unsure of the safety and comfort of this exercise, consult your physician before performing these exercises.

Head rotation

Begin by rotating your head slowly to the right. Notice how far you can rotate your head without strain. Now, rotate your head to the left. Again, notice how far you can rotate your head without strain. If you experience a restriction of motion on either side, you may need to avoid one-sided activities and/or begin stretching your body more often. The goal is to have symmetrical range of motion (Fig. 12-3).

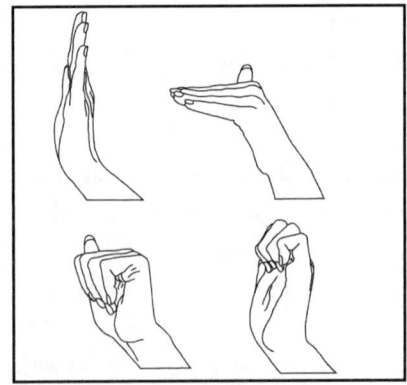

Fig. 12-1 Hand Stretches.

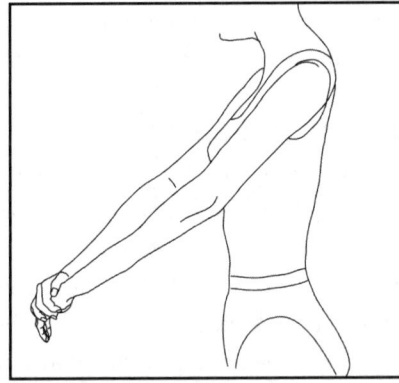

Fig. 12-2 Forearm Stretches.

That's a Stretch!

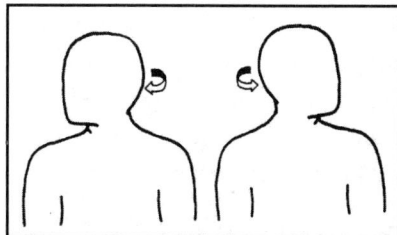

Fig. 12-3 Head Rotation. *Reprinted with permission from Sitting on the Job, Scott Donkin, D.C.*

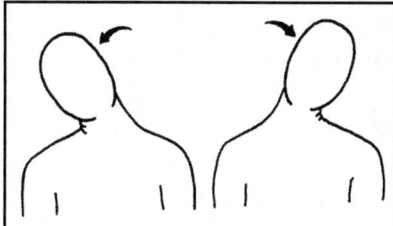

Fig. 12-4 Head Tilt. *Reprinted with permission from Sitting on the Job, Scott Donkin, D.C.*

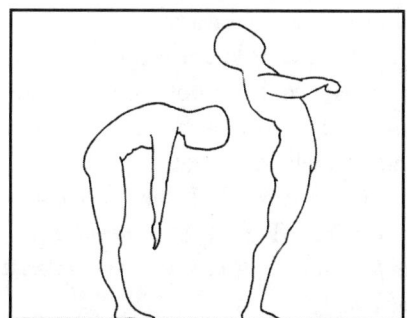

Fig. 12-5 Forward and Back Bending. *Reprinted with permission from Sitting on the Job, Scott Donkin, D.C.*

Head tilt

Focus on an area directly in front of you and tilt your head to the right and then to the left, making sure your head remains forward. Watch for uneven extensions. The objective is to gain even motion on both sides. Look for uneven motion and painful restrictions. Do not force motion beyond a range of ease. You may also tilt your head forward towards your chest and then backward towards your bottom. Again, be mindful of uneven motion and tightness (Fig. 12-4).

Do not force movement or cause painful exertions.

Forward and back bending

Bend your body forward in an effort to touch your toes. Do not force your body beyond a comfortable range. The degree of bending is usually about 90 degrees. Now bend your body backward, approximately 30 degrees. Do not force beyond range of ease (Fig. 12-5).

Side bending

Dental team providers tend to hold their weight on the side closest to the patient. To even out weight distribution and stretch both sides

ALL *the* RIGHT *MOVES*

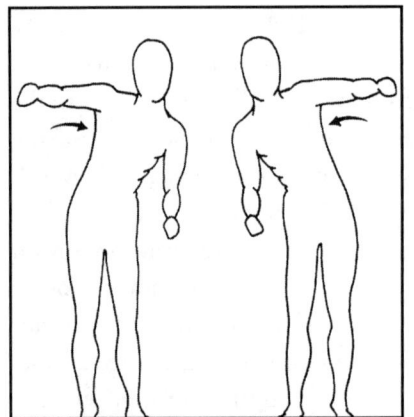

Fig. 12-6 Side Bending. Reprinted with permission from Sitting on the Job, Scott Donkin, D.C.

of the body, bend your torso from side to side, approximately 30 degrees in each direction, without strain (Fig. 12-6). Do not force beyond the range of ease.

Trunk rotation

While standing straight, with feet flat on the floor, rotate your upper body only, at the trunk, to the right approximately 30 degrees or greater. Repeat to the left (Fig. 12-7). *Remember not to force your body beyond your comfortable range.*

Abdominals

The abdominal muscles are among the first muscles to become soft and weak from long-term sitting. The reverse sit-up is a safe and effective intervention. Position your body on a mat lying on your back with your arms at your sides and your palms downward.

Bend your knees and place your feet flat on the floor. Slowly raise your bent legs upward towards your face. Slowly lower your legs into the home position and then do eight repetitions (Fig. 12-8) *Remember not to force your body beyond your comfortable range.*

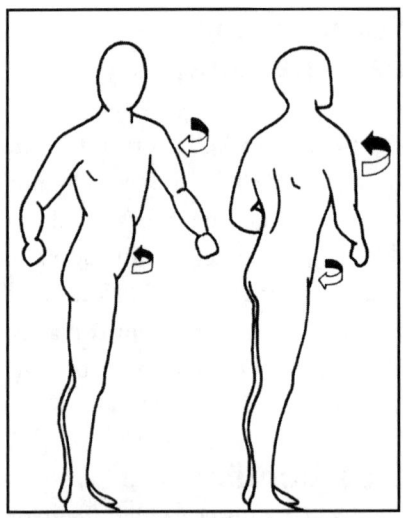

Fig. 12-7 Trunk Rotation. Reprinted with permission from Sitting on the Job, Scott Donkin, D.C.

That's a Stretch!

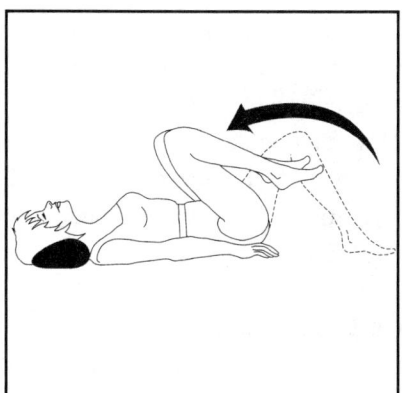

Fig. 12-8 Abdominals.

Buttocks

The buttocks and inner thighs are frequent trouble spots for people who sit, because the buttocks muscles bear the weight of the upper body.

To lighten the load, tuck the pelvis and lift and lower each leg while tightening the buttocks and back thigh muscles.

Be sure your chair is stable enough to support weight shifts before attempting this exercise!

Shoulder shrugging

If you use your arms to hold items such as the suction grip, rotary equipment, or a telephone handset for long periods of time, you may need to stretch frequently.

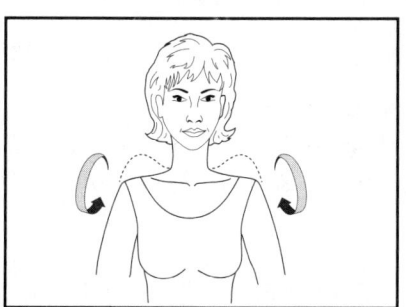

Fig. 12-9 Shoulder Shrugging.

A good exercise for this would be shoulder shrugging. Begin by shrugging your shoulders up toward your ears and then roll them backward. Repeat by rotating them forward in a circular motion (Fig. 12-9).

Mini breaks

Mini breaks can be taken advantage of when you have a lull in your work activities or in between patient appointments.

For those who sit for long periods of time, it is important to stretch your leg muscles by straightening your legs and moving your ankles and feet. Flexing and extending the feet, rotating your ankles clockwise and counter clockwise, lifting and lowering your legs, and stretching your upper body behind you can accomplish this.

ALL *the* RIGHT *MOVES*

Fig. 12-10 Mini Breaks *Taken with permission from Sitting on the Job, Scott Donkin, D.C.*

Again, before attempting this exercise, be sure your chair is stable enough to support you as you shift your weight. This mini-break exercise can refresh your body, and improve comfort, thus supporting an increase in productivity (Fig. 12-10).

Posture check

If you sit for long periods of time, periodically throughout procedures, consciously push your buttocks back into the seat pan of your chair. Next, raise your chest out like a soldier and lift your chin up.

Vision break

To take a break from the lights and computer screen, lightly place the palms of your hands over your eyes while they are closed. Hold for approximately 45 seconds. You may also practice looking in the opposite direction of your work from time to time.

End of a workday

At the end of your workday, choose to do things that help you unwind.

Stress reduction techniques—such as audio relaxation, meditation, or activities like aerobic exercises three to four times a week—can make a positive difference in your health and well-being. You may also want to participate in group sports activities, as they can also be fun and invigorating.

ADDENDUM

The Occupational Safety and Health Administration

Since the Occupational Safety and Health Administration (OSHA) was created in 1970, the agency's mission has been clear and unwavering: "To save lives, prevent injuries and illnesses, and to protect the health of America's workers."

OSHA has 25 state or territorial partners: Alaska, Arizona, California, Connecticut, Hawaii, Indiana, Iowa, Kentucky, Maryland, Michigan, Minnesota, Nevada, New Mexico, New York, North Carolina, Oregon, Puerto Rico, South Carolina, Tennessee, Utah, Vermont, Virgin Islands, Virginia, Washington, and Wyoming.

Both the federal- and state-run programs share tremendous successes in workplace safety. For instance, since 1970, the overall workplace death rate has been cut in half. OSHA's cotton dust standard virtually eliminated brown lung disease in the textile industry; deaths from trench cave-ins declined by 35%; and OSHA's lead standard reduced blood poisoning in battery plants by two-thirds.

ALL *the* RIGHT *MOVES*

OSHA inspections (which are feared by so many) have actually revealed positive results. A post-inspection study showed that in the three years following an OSHA inspection, injuries and illnesses drop on average by 22% per workplace.

OSHA has also found that injury and illness rates have declined in the industries where they have concentrated their attention. Yet, injury and illness rates have remained unchanged or have actually increased in the industries where they have had less or no presence.

Unfortunately, despite OSHA's efforts, every year more than 6,000 Americans die from workplace injuries. An estimated 50,000 people die from illnesses caused by workplace chemical exposures, and another 6 million people suffer non-fatal workplace injuries. This costs the U.S. economy more than $110 million a year. What's really heartbreaking is that the majority of these injuries are predictable and preventable, had employers embraced simple safety awareness principles.

OSHA reform

Businesses often complain about OSHA's overzealous enforcement and burdensome rules. Many people still see OSHA as an agency so enmeshed in its own red tape that it has lost sight of its own mission. To that end, OSHA has agreed to reform the way it does business by decreasing the red tape and paperwork in an effort to keep pace with the workforce and problems of the future. It will, however, hold strong to its mission to increase the protection of worker health and safety.

Above all else, the "new OSHA" will seek to ensure that safety is promoted and protected by those in the workplaces themselves—managers and workers at the worksite—so that they can *interact* with the businesses, rather than react. OSHA claims that it will focus on the most serious hazards and the most dangerous workplaces and insist on results, instead of red tape!

OSHA is claiming that it is not changing direction because it has changed its ultimate destination, but rather embracing change to reach its destination in a better way. That destination—"An America whose workplaces, as far as possible, are free from hazards that are causing or likely to cause death or physical harm"—must be kept in sight at all times. OSHA promises to find and punish employers who neglect safety and endanger their workers with strict, vigorous enforcement.

Addendum

Working together

Since all workplaces are not alike, and not all employers are equally responsible, OSHA plans to take steps to treat employers with aggressive health and safety programs differently from employers who lack such efforts.

In effect, employers will be offered a choice of how they will be regulated by OSHA.

This approach has the potential to dramatically increase safety and ease the adversarial relationship between regulators and business, by putting primary responsibility for ensuring safety in the hands of managers and workers at worksites across the country.

At its core, this new approach seeks to encourage the development of worksite health and safety programs.

With a health and safety program, employers and employees work together to find the best solutions to the particular problems of their workplace. OSHA will be looking for programs with these features:

- management commitment
- meaningful participation of employees
- a systematic effort to find safety and health hazards *whether they are covered by existing standards or not*
- documentation that the identified hazards are fixed
- training for employees and supervisors
- proof that a reduction in injuries and illnesses has occurred

Consultation services

OSHA provides consultation assistance at no cost to employers who request help in establishing and maintaining a safe and healthful workplace.

Comprehensive assistance includes an appraisal of all mechanical physical work practices and environmental hazards of the workplace and all aspects of the employer's present job safety and health program.

At this writing, OSHA's consultation services have been underutilized in dentistry due to distrust of governmental agencies—and as a result of what could potentially end up being self-induced retaliation. Nonetheless, OSHA has provided its consulting services to hundreds of business around the country and reports hundreds of success stories.

One of their successful case studies involved a plastics company in Connecticut which, after OSHA's consultation, was able to cut lost workdays related to on-the-job injuries from 14.2 to 5, and to reduce the number of hazards cited from 138 on the initial visit to only four on the follow-up visit. While these risks are not present in a dental environment, the results are quite impressive.

OSHA's Ergonomic Evolution

After several years of fighting ferocious battles, OSHA published an ergonomic program standard to protect the musculoskeletal health of American workers.

Their mission was to reduce the severity of musculoskeletal disorders (MSD)—disorders involving muscles, nerves, tendons, ligaments, joints, cartilage, blood vessels, or spinal discs. The standard published on November 14, 2000 held an effective date of January 16, 2001, with a deadline date (for implementation) set for October 14, 2001.

On March 6, 2001—less than two months after the standard was released—the Senate (supported by President George W. Bush) repealed the standard that was put in place by former President Clinton just before he left office.

This reversal was made possible by exercising a never-before-used legislative process that expedites repeals without an option for amendment.

Consequently, musculoskeletal safety in the workplace remains voluntary, with the exception of the state of California—the only state in the nation with its own ergonomic program at the time of this publication.

As for the other 49 states, the question remains, "Will they take it upon themselves—regardless of government intervention—to do the right thing?"

The real question to be asked is this: Is it better to be proactive, as opposed to doing nothing at all?

As we explore these two options, allow me to provide some insight and incentive to encourage you to embrace an exciting opportunity to reduce physical harm to workers.

Addendum

Ergonomic principles

Let's start by first examining the mission behind ergonomic principles.

Ergonomic practice is designed to minimize worker fatigue and reduce musculoskeletal stress through balanced neutral postures, time and motion conservation, and providing and maintaining healthy environmental surroundings.

To that end, when employees are forced to operate beyond their physical capacity or suffer a mismatch between physical capacities and the physical requirements of the job, poor performance and accidents are more likely to occur. Depending on the severity of the stress or injury, the business becomes more vulnerable for lost workdays and additional costs in hiring and training employee replacements—this, notwithstanding the risk of increasing medical costs, worker's compensation, and insurance costs.

While most individuals do not choose to see these unfortunate consequences as real potential risks, the mere opportunity to increase employee performance and levels of productivity should be compelling enough to *do the right thing!*

Doing the right thing

If you desire a reduction in work-related risks and an increase in production potential, you may choose to use the following abbreviated guidelines (taken from OSHA's ergonomic standard) as an outline for developing your own voluntary program.

[For readers in those states that have developed their own ergonomic programs, such as California, be sure to follow your specific state plan.]

First, we suggest that you kick-start your efforts by simply providing your team with basic information on musculoskeletal safety and health.

It can be accessed from *What you Need to Know About Musculoskeletal Disorders (MSDs)* and *Summary of the OSHA Ergonomics Program Standard* available at—

www.osha-slc.gov/ergonomics-standard/regulatory/AppendixA

and

www.osha-slc.gov/ergonomics-standard/regulatory/AppendixB

The next step involves training employees to understand the potential risks related to their specific job tasks. Each employee must also be able to recognize signs and symptoms of MSDs as well as understand the process of reporting signs and symptoms upon their onset.

Reporting

When an employee reports MSD signs or symptoms, it is best to first determine if the employee's signs and symptoms constitute an "MSD incident." A work-related incident results in days away from work, restricted work, or medical treatment beyond first aid, for symptoms that last for seven consecutive days after the initial report. When it is determined that the MSD is an MSD incident, the job should be screened to determine if repetition, force, awkward postures, contact stress, or vibration risk factors exist.

Triggers

If a job involves these types of risks on one or more days a week, the MSD "incident," according to OSHA, has become an "action trigger." If a job meets the requirements of an action trigger, the business owner can "quick-fix" the problem.

A quick-fix is understood to be a solution to a problem that can be implemented within 90 days, providing that in the history of that job, only one prior MSD incident in that same job and only two prior MSD incidents in the entire establishment have been recorded within the past 18 months.

Work restriction

Ideally, the employee should be provided work restriction protection (WRP) when he/she reports MSD signs or symptoms—such as pain, numbness, tingling, burning, cramping, and stiffness—as well as a decreased range of motion, deformity, decreased grip strength, and loss of muscle function.

If the sign or symptom is subsequently determined to be a work-related action trigger, then the employee is restricted from that work-related action and is granted a WRP (as determined by the supervising healthcare professional and employer) for as long as necessary to ensure recovery.

What risks are present in dentistry?

In dentistry, poor postures are very common. Patient comfort tends to be much more of a concern than the comfort of the operator or dental team

Addendum

treating the patient. Unfortunately, poor posture can, over time, cause a plethora of MSDs, and should be monitored to ensure musculoskeletal health and comfort.

Of equal importance, ergonomically designed equipment, technology, and instruments can minimize MSD risk factors. For example, operator stools are designed to encourage neutral seated positions and offer adjustable armrests, improved lumbar support, and more options for adjustment. Patient chairs feature tapered backs to allow team members to sit closer to the patient, minimizing the need to lean forward. Hand-specific gloves have also been shown to exert less stress on the hands than ambidextrous styles. New computer workstation designs can also accommodate visual access and reduce stress from repetitive motion.

Proactive ergonomic approaches are said to prevent more than 300,000 MSDs each year. If we ignore the signs and symptoms and continue to wait for the governmental bureaucracy to take charge, a predictable path towards early retirement may become a reality—making the emotional pain in leaving the profession perhaps even greater than the actual physical pain suffered on the job!

You can provide your dental workers the opportunity to work smarter, *not harder*, and proactively prevent common musculoskeletal and nervous system disorders. Don't wait until these guidelines are made mandatory by an outside entity. As Mark Twain said, "Twenty years from now you'll be more disappointed with the things you didn't do, rather than the ones you did." Let there be no regrets.

Begin by observing your habits and choose to conscientiously move towards greater efficiency, balance, and comfort. There is great promise in the many rewards gained through ergonomic awareness. Businesses that encourage a health-smart practice life-style and that embrace a healthier work-style for themselves and their team partners will make a good choice of doing the right thing, because the right thing is a good thing!

BIBLIOGRAPHY

1. *Building Air Quality (BAQ)*, NIOSH/EPA, Department of Health & Human Services Publication

2. Donkin, S., *Sitting on the Job*, Houghton Mifflin Company, Boston, 1986

3. *Ergonomic Workstation Set-Up*, Broderbund, Novato, CA, 1997, CD-ROM product

4. "Ergonomics & The Dental Hygienist At Work," *Access*, January 1995

5. Federal OSHA Ergonomic Draft, March 13, 1995

6. Federal OSHA Ergonomic Draft, May 3, 1999

7. Garcia, Arturo R., "The Benefits of High Magnification," *Dental Economics*, June 2001

8. Grandjean, E., *Fitting the Task to the Man*, Taylor & Francis, London, 1980

9. *Healthcare Facility Environmental Floor Coverings,* Collins & Aikman Floorcoverings, Inc., 1998

10. Lawreance, G., "Dental Office Massage Therapy," *The Journal of Practical Dental Hygiene,* March-April 2000

11. Leisemeyer, G., *Design Standards,* Sullivan-Schein Manual, 1998

12. Murphy, D., *Ergonomics and the Dental Care Worker,* APHA, 1998

13. *Occupational Health Supplement,* National Health Interview Survey, Department of Health & Human Services, 1993

14. Occupational Safety and Health Administration, "Final Ergonomics Program Standards," *Federal Register,* Volume 65, No. 220: 68262-68870 (November 14, 2000) www.simonsaysseminars.com.oshasic/ergonomics-standard/regulatory/regtext.html

15. Occupational Safety and Health Administration, *Mission Statement* www.OSHA.gov/oshinfo/mission.html

16. Poindexter, S. Marshall, "All the Right Moves," Access, January 1995

17. Pollack, R., "Dento-Ergonomics," *CDA Journal,* April 1996

18. Pollack, R., "Environmental Design Promotes Compliance," *Dentistry Today,* 1993

19. Pollack, R., "Ergonomics," *Dental Teamwork,* September-October 1989

20. Pollack, R., *Interactive Ergonomics,* Network One, 1995, video

21. Pollack-Simon, R., *Conquering Compliance,* Simon Says Seminars, inc., 2000 (chapter 12)

22. Pollack-Simon, R., *Smooth & Efficient Patient Care @ the Chair,* Simon Says Seminars, inc., 1999

23. *Principles of Ergonomics,* ErgoWeb, www.ergoweb.com, 1998

Bibliography

24. Rucker, L. et al., "Declination Angle and Its Role in Selecting Surgical Telescopes," *JADA,* July 1999

25. Sieg and Adams, *Illustrated Essentials for Musculoskeletal System,* 1997

26. Stitik, T., "An Analysis of Cumulative Trauma Disorders in Dental Hygienists," *The Journal of Practical Hygiene,* March-April 2000

27. "The Anthos Chair Delivery System," *Dental Town Magazine,* January 2002, page 26

28. *The Selection Process,* Koroseal Wallcoverings, Fairlawn, Ohio, 1977

29. *20 Questions About Fitness,* The Firm, BMG, video

30. U.S. Department of Labor, Office of Public Affairs, OSHA National News Release, March 24, 2000

31. *Ventilation and Air Quality in Offices,* EPA Office of Air and Radiation, Fact Sheet (6607J), 402-F-94-003

32. *What is Low Pressure Melamine?,* Alpine Plywood Corporation, 1998

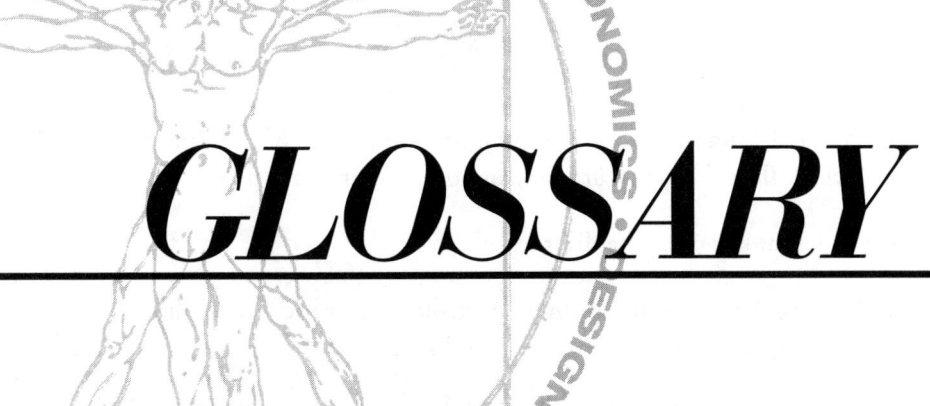

GLOSSARY

AB Switch. A mechanism that allows a piece of equipment to switch from two separate sources (e.g., a TV monitor from television to VCR).

Ambidextrous gloves. Gloves designed to fit on both right and left hands. Gloves that are not fitted for a specific hand and are molded in a flat position.

ADA. American Dental Association.

Abduction. A limb or digit moving away from the midline of the body.

Adduction. A limb or digit moving toward the midline of the body.

ASHRAE. American Society of Heating, Refrigerating, and Air Conditioning Engineers.

BP. Blood pressure.

CPU. Central processing unit.

ALL *the* RIGHT *MOVES*

CT (carpal tunnel). The pathway on the palmar side of the hand through which flexor tendons, arteries, and the median nerve pass.

CTD. Cumulative trauma disorder.

CTS (carpal tunnel syndrome). A result of constriction to the median nerve resulting in tingling, pain, and numbness.

DAU (Dental Auxiliary Utilization). Utilizing trained individuals to assist the clinician in an effort to ease time and motion on behalf of the clinician.

Ergonomics. The application of human performance abilities and limitations related to equipment, tools, and environments. The science of fitting the job to the worker in an effort to minimize musculoskeletal disorders.

Electronic air cleaners. Devices that use an electronic charge to remove airborne particles.

Ergosize. A term to describe the sizing of work environments as it relates to ergonomic principles.

HEPA. High-efficiency particulate air.

HVAC. Heating, ventilation, and air conditioning.

HVE. High-velocity evacuator.

Hypothenar eminence. Fleshy elevation on palmar side of hand.

LPM. Low-pressure Melamine.

MSD. Musculoskeletal disorder.

Musculoskeletal disorders. An injury to or a disease of the muscles or bones of the body, typically work related.

NFPA. National Fire Prevention Association.

NIOSH. National Institute for Occupational Safety and Health.

Glossary

OSHA. Occupational Safety and Health Administration.

Patient viewing monitors. Monitors used for patient education and image presentation. These monitors can be digital or analog.

RMI (repetitive motion injury). An injury that results from repeated movements or motions, usually involving forceful exertions, deviations, neutral positions, and postures, without safeguards or administrative controls, such as breaks.

SBS. Sick building syndrome.

SPI. Stitches per square inch.

VDT (visual display terminal). Monitors used to view images and text from computers.

WRP. Work restriction protection.

RESOURCES

American Dental Technologies	Anthos modular high-tech delivery system	877 793-3717 www.americandentaltech.com
CASEY Education Systems, Inc.	Patient education, continuing ed, and website development	800 683-5409 www.casey.com
Cieos	CIEOPORT: compact computer with LCD, flat-panel based touchscreen	800 627-3205 www.cieos.com
Donkin, S. W.	*Sitting on the Job* Houghton Mifflin Co.	Donkin007@aol.com 800 552-6347
DCI/ Pelton Crane	Ergonomic dental units	800 659-3212 www.dcionline.com
Gendex/Densply	AcuCam Concept IV digital camera	800 800-2888 www.gendexxray.com
Global Surgical	Microscopes, micro surgeon	800 767-8726
Health by Design	Bambach Saddle Seat	888 909-ERGO www.healthbydesign.com

ALL *the* RIGHT *MOVES*

Hello Direct	Phone headsets	800 444-3556 www.hellodirect.com
Hu Friedy	Ergonomic instrument designs, instrument cassette systems	800 729-3743 www.hufriedy.com
Link Ergonomics	Assistant stools 185 ABR	800 424-5465 www.linkdental.com
Newport Coast Oral Facial Institute	Multidisciplinary hands-on mentoring	800 686-1155 www.ncofi.org
Natures Intense Therapy	Joint rub for hands and muscles	888 577-6972
OSAP	Organization for Safety and Asepsis	800 298-6727 www.osap.com
OSHA	Occupational Safety & Health Administration	202 219-8151 www.osha.gov
Pillar Design Group	Office planning and design	262 363-0575 pillar.design@voyager.net
RGP	Operator stools with telescoping armrests	800 522-9695 www.rgpdental.com
Samsung	Flat screen monitors	877 793-3717 www.samsung.com
Simon Says Seminars, Inc.	Team coaching, seminars and video library	800 366-8326 www.simonsaysseminars.com
Smart Practice	Gloves right/left	800 522-0800 www.smarthealth.com
SurgiTel/ General Scientific	Magnification and illumination	800 959-0153 www.surgitel.com
Unthank Design Group	Architectural and design firm	402 423-3300 www.unthank.com

About the Author

Risa Pollack-Simon is a certified management consultant, nationally recognized speaker, published author, and producer of video and audiotape programs that have been recognized for continuing education by the American Dental Association. Risa is the president of Simon Says Seminars, Inc.—a practice management company located in Scottsdale, AZ. Since 1982, Risa has been coaching dental teams across America to provide extraordinary "service," in ergonomically efficient and highly productive, high-tech, patient care environments.

Risa's memberships include the National Speakers Association and the Institute of Management Consultants USA, where Risa obtained the award of certification. This certification represents evidence that Risa meets the highest standards of ethics in management consulting—an achievement represented by only 1% of the entire consulting profession nationwide. Risa is also a member of the American Association of Women Dentists; the Organization of Safety & Asepsis Procedures; and the Academy of Dental Management Consultants. Additionally, Risa is a Consultant to the ADA's

ALL *the* **RIGHT** *MOVES*

Council on Dental Practice, and served as the editor for the ADA's magazine *Dental Teamwork*. For more information contact:

Simon Says Seminars, Inc.
34522 N. Scottsdale Road PMB #623
Scottsdale, Arizona 85262
800 366-8326 Voice
480 575 9357 FAX
www.simonsaysseminars.com
risa@simonsaysseminars.com

INDEX

A

Abdominal support chair, 44: rationale, 44; recommendation, 44

Abdominals (exercise), 84-85

Accessibility, 28-32, 34, 39-40, 44, 46-47, 72-75: delivery system, 39-40; chair controls, 46-47

Acronyms, 99-101

Action trigger, 92

Administrative controls, 11, 13

Administrative workspace, 49-52: workstation design, 49; keyboard, 49-50; mouse, 50; screen, 50-51; chair, 51; desktop, 51; telephones, 52

Air quality/ventilation, 62, 66-68

American Society of Heating, Refrigerating and Air Conditioning Engineers (ASHRAE), 67

Angle of declination, 22

Armamentarium, 39-42

Attitude, 2

Australian saddle-style chair, 19-20

Author information, 105-106

Automation, 55-56: practice management monitors/keyboards, 55-56; extension arms, 55; cordless keyboards, 55; touch-screen monitors, 55; light-pen operation, 55; voice activation, 55; patient viewing monitors, 55; X-ray units, 55-56

ALL *the* RIGHT *MOVES*

B

Balance/imbalance (body), 4-5
Behavior modification, 12-13
Bending, 83-84: forward and back, 83; side, 83-84
Bibliography, 95-96
Biomechanics, 19-20
Body parts and mechanics, 3-9, 19-20: checks and balances, 4-5; sculpting, 5; move and groove, 5; pain, 6; hand/wrist disorders, 6-7; other factors, 7-9; rotator cuff tendonitis, 8; lateral epicondylitis, 8; medial epicondylitis, 8; de Quervain's tenosynovitis, 9; massage therapy, 9
Body position (team/individual), 27-32, 72-75: patient positioning, 31-32; maxillary position, 31-32; mandibular position, 32
Body sculpting, 5
Buttocks, 85

C

Cabinetry, 42-43, 56-59, 62-64
Carbon dioxide, 66
Cardio-respiratory fitness, 79-80
Carpal tunnel syndrome, 6-7
Cassette instrument management system, 60-61
Central processing unit, 72-73
Chair back thickness, 45: rationale, 45; recommendation, 45
Chair controls accessibility, 46-47: rationale, 46; recommendation, 46-47
Chair design, 4-5, 15-20, 31-32, 44-47, 51, 76
Chair positioning, 31-32
Chair support, 17-19
Checks and balances (body), 4-5
Clinical equipment, 37-48: major design flaws, 39-47; enhancements, 46-48; workspace, 48
Coiled handpiece tubing, 45-46: rationale, 46; recommendation, 46
Comfort factors (networks design), 75-77
Compressor equipment, 65
Consultation services (OSHA), 89-90
Control equipment, 72-75
Control strategy, 11-13
Convergence angle, 22, 47
Coordination, 7
Cordless keyboards, 55
Counterbalance, 4
Cross-training, 70
Cumulative trauma disorder, 15
Cuspidor attachment, 42-44: rationale, 42-43; recommendation, 43-44

Index

D

Darkroom, 63-64

Delivery system accessibility, 39-40: rationale, 39-40; recommendation, 40

Delivery system, 39-42, 54-55: accessibility, 39-40

Depth of field, 21-23

DeQuervain's tenosynovitis, 9

Design (optics/visual aids), 23-25: through-the-lens, 23; flip-up style, 23-24; microscopes, 23-25

Design (workspace), 49-52: workstation, 49; keyboard, 49-50; mouse, 50; screen, 50-51; chair, 51; desktop, 51; telephones, 52

Design plan. SEE Master design plan AND Networks design plan.

Design/equipment layout flaws, 39-47: delivery system accessibility, 39-40; support equipment mounting, 40-42; work surface space, 41-43; cuspidor attachment, 42-44; abdominal support chair, 44; chair back thickness, 45; coiled handpiece tubing, 45-46; chair controls accessibility, 46-47

Desktop, 51

Digital devices, 71

Direct lamp lighting, 24-26

Discs (spinal), 3

Dispensers, 59

Doing the right thing, 91-92

E

Electronic charting, 70, 73

Engineering controls, 11-12

Environmental space factors, 33-35, 41-43, 48: job station analysis, 34-35; strategic plan, 35; lighting, 35; noise, 35; workspace, 41-43, 48

Epicondylitis, 8

Equipment layout, 12-13, 37-48

Ergonomic awareness, ix-x, 1-2, 11-13, 33-35, 53-54, 69-77, 90-93: attitude, 2; stress management, 2

Ergonomic evolution (OSHA), 90-93: ergonomic principles, 91; doing the right thing, 91-92; reporting, 92; triggers, 92; work restriction, 92; risks in dentistry, 92-93

Ergonomic practices, 11-13

Ergonomic principles, 91

Ergonomics history, 53-54

Exchanges, 29-30

Exercises (stretching), 79-86: health and fitness, 79-81; cardio-respiratory fitness, 79-80; muscle fitness, 80-81; stretching warm-up, 81-86; hand stretches, 81-82; forearm stretches, 82; head rotation, 82-83; head tilt, 83; forward and back bending, 83; side bending, 83-84; trunk rotation, 84; abdominals, 84-85; buttocks, 85; shoulder shrug-

ging, 85; mini breaks, 85-86; posture check, 86; vision break, 86; end of workday, 86
    Extension arms, 55

F

Fatigue, 5-6, 75-77
Fiber optics, 24-26
Finklestein test, 9
Fitness/health, 79-81: cardio-respiratory fitness, 79-80; muscle fitness, 80-81
    Flexor retinaculum, 7
    Flip-up style (optics), 23-24
    Floor coverings, 59-60, 63
    Forearm stretches, 82
    Forward and back bending, 83
    Four-handed dentistry, 16-17
    Future technology, 69-77: modernization, 69-70; networks, 71-77

G

Glossary, 99-101
Gloves, 7
Golfer's elbow, 8

H

Halogen light, 24-26
Hand stretches, 81-82
Hand/wrist disorders, 6-7
Handpieces, 12, 27-30, 34-35, 39-42
Hazards, 1, 7, 9, 87-93
Head rotation, 82-83
Head tilt, 83
Healing, 6, 9
Health and fitness, ix, 53, 79-81, 87-93: cardio-respiratory fitness, 79-80; muscle fitness, 80-81
High velocity evacuator (HVE), 28, 43-44
High-efficiency particulate air (HEPA) filters, 67-68
History (design), 53-54
HVAC system, 66-68
Hygiene, 57-62

I

Illumination, 24-26, 47
Incident (MSD), 92
Inflammation, 6, 8-9
Information resources, 103-104
Instrument design, 7
Instrument recirculation, 60-61
Integrating technology, 69-77: modernization, 69-70; network design, 71-77; investment, 76-77
Intervention/prevention, 11-13: engineering controls, 12; work practice controls, 12-13; administrative controls, 13
Investment in ergonomics, 69-70, 76-77

Index

J

Job station, 34-35, 49: analysis, 34-35
Joints, 5

K

Keyboard, 49-50, 55-56, 76

L

Laboratory, 62-63
Lateral epicondylitis, 8
Laundry, 64-65
Legionnaire's disease, 66
Legroom, 35, 45, 51
Ligaments, 5
Light intensity, 24-26
Lighting/light source, 24-26, 34-35
Light-pen operation, 55
Limbs, 4
Lordosis, 9
Loupes, 21, 23, 47
Low-pressure Melamine®, 56-57, 61

M

Magnification, 21-26, 46-48: design, 23-25; through-the-lens, 23; flip-up style, 23-24; microscopes, 23-25; illumination, 24-26
Mandibular position, 32
Massage therapy, 9
Master design plan, 53-68: history, 53-54; treatment room environment, 54-55; automation, 55-56; cabinetry, 56-59; dispensers, 59; floor coverings, 59-60; instrument recirculation, 60-61; sinks and ultrasonics, 61-62; ventilation, 62, 66-68; laboratory, 62-63; darkroom, 63-64; public traffic areas, 64; restrooms (staff/doctor), 64; laundry, 65; mechanical equipment rooms, 65; nitrous oxide storage, 65-66; ventilation and air quality, 66-68
Maxillary position, 31-32
Mechanical equipment rooms, 65
Medial epicondylitis, 8
Microscopes, 21, 23-25
Mini breaks, 85-86
Modernization/technology, 69-70
Monitors, 55-56
Mouse (computer), 50
Move and groove, 5
Muscles, 3, 17-18, 80-81: fitness, 80-81
Musculoskeletal disorder, ix-x, 1, 11-13, 33-35, 38, 75, 90-93

N

National Fire Prevention Association (NFPA), 66
National Institute for Occupational Safety and Health (NIOSH), 37, 54
Network installation, 71
Networks design plan, 71-77: working smarter, 75; visual display comfort, 76; keyboard comfort, 76; chair comfort, 76; securing the future, 76-77
Neutral posture, 15
Nitrous oxide storage, 65-66
Noise, 34-35

O

Occupational health hazard, ix, 1, 7, 9, 87-93
Occupational Safety and Health Administration (OSHA), ix, 38-39, 54, 87-93: OSHA reform, 88; working together, 89; consultation services, 89-90; OSHA's ergonomic evolution, 90-93; ergonomic principles, 91; doing the right thing, 91-92; reporting, 92; triggers, 92; work restriction, 92; risks in dentistry, 92-93
Open-leg position, 19-20
Optics/visual aids, 21-26, 46-48: design, 23-25; through-the-lens, 23; flip-up style, 23-24; microscopes, 23-25; illumination, 24-26
OSHA. SEE Occupational Safety and Health Administration.
Ozone, 68

P

Pain, 6, 8-9, 19
Paresthesia, 7
Patient positioning, 31-32, 34, 72-75: maxillary position, 31-32; mandibular position, 32
Patient viewing monitors, 55
Pinch posture, 7
Positioning (team/individual), 27-32, 34, 72-75: patient positioning, 31-32, 34; maxillary position, 31-32; mandibular position, 32; zones, 72-75
Positioning zones, 72-75
Posture check, 86
Posture, 1, 5, 15-20, 27-34, 40-41, 86: sitting on the job, 16-17; spinal balance, 17-19; sitting posture concepts, 19-20; posture check, 86. SEE ALSO Team/individual positioning.
Practice management monitors/keyboards, 55-56
Prevention and intervention, 11-13: engineering controls, 12; work practice controls, 12-13;

Index

administrative controls, 13
Public traffic areas, 64

Q

Quick-fix solution, 92

R

Reform (OSHA), 88
Regulation/regulatory agency, 87-93
Repetitive motion, 7, 13, 33
Reporting (OSHA), 92
Resources/information, 103-104
Restrooms (staff/doctor), 64
Retraction, 29-30
Right-sizing (workspace), 68
Risk factors, 1, 7, 9, 11-13, 15-16, 33-35, 38-39, 75, 92-93
Risks in dentistry, 1, 7, 9, 92-93
Rotator cuff tendonitis, 8

S

Safety, 87-93
Saliva evacuation, 42-44
Scoliosis, 9
Screen, 50-51
Sculpting, 5
Securing the future, 76-77
Shoulder shrugging, 85
Sick building syndrome, 66
Side bending, 83-84
Single-entry charting, 70
Sinks and ultrasonics, 61-62
Sit-down dentistry, 16-17
Sitting posture, 16-17, 19-20
Spinal balance, 17-19
Spinal nerves, 3
Sterilization, 60-62
Strategic plan, 35: lighting, 35; noise, 35
Stress management, ix-x, 2, 5-6, 9, 16, 33-35, 38
Stretching exercises, 79-86: health and fitness, 79-81; cardio-respiratory fitness, 79-80; muscle fitness, 80-81; stretching warm-up, 81-86; hand stretches, 81-82; forearm stretches, 82; head rotation, 82-83; head tilt, 83; forward and back bending, 83; side bending, 83-84; trunk rotation, 84; abdominals, 84-85; buttocks, 85; shoulder shrugging, 85; mini breaks, 85-86; posture check, 86; vision break, 86; end of workday, 86
Subluxation, 4
Support equipment mounting, 40-42: rationale, 40-41; recommendation, 41
Surgery, 7
Symptoms, 5-6, 8-9, 11

T

Task analysis, 11-13

Team/individual positioning, 27-32: patient positioning, 31-32; maxillary position, 31-32; mandibular position, 32. SEE ALSO Posture.

Technology integration, 69-77: modernization, 69-70; networks, 71-77

Telephones, 52

Telescopes, 21

Temperature, 34

Tendonitis, 8

Tendons, 3, 8-9

Tennis elbow, 8

Tenosynovitis, 9

Terminology, 99-101

Through-the-lens (optics), 23

Time management, 2, 13, 85-86

Tissue memory, 5

Touch-screen monitors, 55

Treatment room environment, 54-55

Triggers, 92

Trunk rotation, 84

U

U.S. Department of Labor, 38

Ultrasonic units, 61-62

V

Vacuum equipment, 65

Ventilation and air quality, 62-64, 66-68: ventilation rate, 68

Ventilation rate, 68

Vertebrae, 3

Video card, 72-74

Vision break, 86

Visual acuity, 21

Visual aids/optics, 21-26: design, 23-25; through-the-lens, 23; flip-up style, 23-24; microscopes, 23-25; illumination, 24-26

Visual display comfort, 76

Visual display terminal, 49-51, 76

Voice activation, 55

W

Warm-up stretching/exercises, 81-86

Waste drops/disposal, 58, 62

Weakness, 7-8

Work practice controls, 11-13

Work restriction protection, 92

Work surface space, 41-43: rationale, 41-42; recommendation, 42

Work surface, 28, 34, 41-43: space, 41-43

Workday end, 86

Index

Working length/distance, 21-22
Working smarter, 75
Working together (with OSHA), 89
Workspace (administrative), 49-52: workstation design, 49; keyboard, 49-50; mouse, 50; screen, 50-51; chair, 51; desktop, 51; telephones, 52
Workspace, 33-35, 41-43, 48-52: job station analysis, 34-35; strategic plan, 35; lighting, 35; noise, 35; administrative, 49-52
Workstation, 34-35, 49: analysis, 34-35; design, 49
Wrist/hand disorders, 6-7

X

Xenon light, 24, 26
X-ray units, 55-56

Z

Zone (positioning), 72-75

Pearls for Your Practice

About the book

At last! After many years as a favorite column in PennWell's *Dental Economics* magazine, Dr. Joe Blaes has collected the very best of his "Pearls for Your Practice" into one comprehensive resource!

Inside this book you'll find preeminent advice about selecting and using not just tools and equipment but services and techniques that will help you and your dental team work smarter and more efficiently — pleasing your patients and your bottom line!

About the author

Joseph Blaes, DDS, has a general practice in the St. Louis area. He is known for his expertise in dental materials and techniques, and his innovative systems designs. Dr. Blaes teaches the behavioral and practice management principles that are so necessary to thrive in a fee-for-service practice. He pioneered many innovative techniques in cosmetic dentistry, and continues to teach these faster, better, easier methods through his lectures and seminars, hands-on workshops, books, videos, and his monthly editorial contributions to *Dental Economics* magazine.

225 Pages/Fall 2001
• $29.95* US • $44.95* INTL
ISBN 0-87814-820-5

*Plus shipping & handling

What's inside

- Materials, Equipment, and Supplies
- Instruments
- Chairside Amenities & Patient Home Care
- Crowns & Bridges
- Dentures
- Bonding
- Restorative/Cosmetic Dentistry
- Services
- Prophys, Sealants, Fillings
- Anesthesia
- Laboratory

3 Easy Ways to Order

Phone: 1.800.752.9764 or +1.918.831.9421
Fax: 1.877.218.1348 or +1.918.831.9555
Website: www.pennwell-store.com

Dental Economics

Because You're a Businessperson Who Happens to be a Dentist!

Business, Legal, and Tax Planning for the Dental Practice, 2nd Edition — **A Must-Have from PennWell Books & More!**

Most of the problems, hurdles, and obstacles that dentists face can be avoided by planning. The second edition of this best-selling title has been completely rewritten and expanded to address considerations for general and specialty practices relating to practice realities of 2000 and beyond — practice transitions, operations, and personnel planning. Business, Legal, & Tax Planning for the Dental Practice, 2nd Edition, identifies opportunities, problems, and decisions which dentists and specialists must make as practice owners and provides solutions. Learn how to positively impact the bottom line by maximizing compensation in all forms, minimizing estate and income taxes, reducing exposure to the legal risks of practice ownership, attaining financial independence, and avoiding poor business decisions. This book takes into account tax law changes passed in June 2001 regarding pensions and estate tax.

While intended for the dentist and dental specialist, this book also is valuable for the doctor's spouse and advisors.

A Comprehensive Guide for Every Dentist
- Associate Buy-Ins and Owner Buy-Outs
- Establishing or Relocating a Practice
- Sale and Purchase of a Practice
- Facility Lease, Purchase or Sale
- Pension and Estate Planning
- Strategic Practice Planning
- Valuation of a Practice
- Staffing Issues
- Fringe Benefits
- Business Deductions

About the Author
For 11 years, William P. Prescott has advised clients on the topics presented in this book. He knows the business, legal, and tax issues facing the dental or specialty practitioner by experience — not theory.

"The information in this book is invaluable for every dentist to have, and it should be on the shelf in its updated version in every dental office in America... I recommend this book to every practicing dentist..."
--from the Foreword by Dr. James R. Pride

Order Details
ISBN 0-87814-800-0 • 480 Pages • $54.95 US plus S&H • 6x9 • Hardcover

Key code: DBAD04

www.pennwell-store.com

Dental Economics

3 Easy Ways to Order: 1. Tel: 1.800.752.9764 or +1.918.831.9421 •
2. Fax: 1.877.218.1348 or +1.918.831.9455 • 3. Online: www.pennwell-store.com

HERE'S WHAT CUSTOMERS ARE SAYING ABOUT SHOPPING ONLINE AT WWW.PENNWELL-STORE.COM:

"The service was great; I had my order within a few days — when all other stores didn't have it in stock."
— Scott R., Accokeek, MD

"I was very pleased with the service. Excellent response to my e-mail inquiring about my order status. I will be ordering from PennWell again in the near future."
— Chester G., Wilmington, DE

"I couldn't find a couple of items, I left an email, and they shipped the items as well. The online store is excellent and has my highest regards and approval."
— Scott E., Ilion, NY

"Being that I haven't ordered online at all in the past, the only basis I had for the quality and speed of service was the feedback from friends and relatives. PennWell has certainly made my first online experience a pleasant one…"
— Hercules R., Westminster, CA

"Already received the order and the invoice — it was quite user-friendly. Will definitely order again online. Thank you!"
— Brenda P., Denver, CO

What are you waiting for? Shop online today at
www.pennwell-store.com!

Don't forget to sign up for our e-newsletter to keep up with our latest titles and offers!